EDUCATION LIBRARY
UNIVERSITY OF KENTUCKY

Race, Rigor, and Selectivity
in U.S. Engineering

RACE, RIGOR, AND SELECTIVITY IN U.S. ENGINEERING

The History of an Occupational Color Line

AMY E. SLATON

HARVARD UNIVERSITY PRESS
Cambridge, Massachusetts
London, England
2010

Copyright © 2010 by the President and Fellows of Harvard College
All rights reserved
Printed in the United States of America

Library of Congress Cataloging-in-Publication Data

Slaton, Amy E., 1957–
Race, rigor, and selectivity in U.S. engineering : the history
of an occupational color line / Amy E. Slaton.
p. cm.
Includes bibliographical references and index.
ISBN 978-0-674-03619-2 (alk. paper)
1. Engineering—Study and teaching—United States—History. 2. Discrimination in
education—United States—History. I. Title.
T73.S487   2010
620.0071'173—dc22   2009024813

Contents

Preface   vii

List of Abbreviations   xiii

1 Introduction   1

2 Identity and Uplift: Engineering in the University of Maryland System in the Era of Segregation   19

3 The Disunity of Technical Knowledge: Constructions of Racial Difference in Separate but Equal Engineering Education   48

4 Opportunity in the City: Engineering Education in Chicago, 1960–1980   79

5 Urban Engineering and the Conservative Impulse: Research at the University of Illinois at Chicago and the Illinois Institute of Technology   113

6 Race and the New Meritocracy: Engineering Education in the Texas A&M System, 1980 to the Present   143

7 Standards and the "Problem" of Affirmative Action: Departures from Convention in the TAMU System   171

8 Conclusion   205

Notes   221

Index   271

Preface

THIS BOOK reached completion in the months following the historic 2008 national election of Barack Obama, the first African American president of the United States. This momentous departure from years of racial tension and inequity in the United States promises immense changes in government and corporate policies and in everyday interactions among individuals. The nation seems poised to ask badly needed questions about the meaning of race, growing economic disparities among Americans, and our failure to achieve the democratic aims so vibrantly declared in the civil rights era. This attitudinal shift does not in itself guarantee a reversal of every discriminatory habit in the nation's educational and hiring structures or in the distribution of housing, health care, and economic opportunity. It may make those changes newly viable, but it has not rendered race immaterial. Instead, we are perhaps willing to acknowledge that race has been a factor in our public and private experiences of the past six decades. For the tremendous promises of this election to reach fruition, we need to take advantage of these new ideological commitments and energy and to look critically at the practices that have sustained racial discrimination for decades beyond the Civil Rights and Voting Rights acts of the mid-1960s, which supposedly assured a new direction in American race relations.

This book turns our attention to the role that race has played in one area of American work and life rarely analyzed from this perspective: engineering. It centers on engineering as it has developed as a field of higher education since the 1940s, a field that often undertook significant reforms meant to increase its racial diversity but that in some instances backed away from those projects as the national climate surrounding race reform efforts cooled. It describes the work of dedicated educators and policy makers and the considerable political and cultural obstacles they faced in their efforts to correct the underrepresentation of blacks in technical occupations. Those obstacles did not often arise from what one could easily label as racist impulses. Many of the actors of importance to this story were not even thinking about race in any direct way; rather, it was in striving to eradicate such thoughts and instead function *as a meritocracy* that higher technical education often reasserted inequitable ideologies. We might now be at a point that the dismissal of affirmative action programs in higher education, which emerged in the 1990s, might be rethought, and I hope to enable that project with this book. Through a series of case studies comparing understandings of race in historically black and predominantly white engineering schools, I survey what has gone before and where we might now go in what can only be called a new era in American civic life.

In 2007, even as the first stirrings of today's electoral transformations were taking shape, a flurry of disturbing events occurred on U.S. high school and college campuses, and these may illustrate where some of the challenges ahead may lie. The display of racist sentiments and heated conflicts about the legal treatment of the "Jena Six," a group of African American students in Louisiana who had responded to the hanging of a noose on a schoolyard tree, caught the imagination of other young people around the country. In the course of a few weeks, an African American professor at Columbia University found a noose on her office door, and similar events occurred at the University of Maryland, Central Michigan University, and at high schools in North and South Carolina. At my own school, Drexel University in Philadelphia, someone scribbled a racial epithet across an Africana Studies hallway bulletin board. In each case, debates flared about whether these were "meaningful" indicators of persistent racial tension in the United States or merely expressions of youthful folly that had unfortunately become fodder for sensationalist media. New York City held a day of rallies against hate crimes as the controversy heightened; however, critics claimed that attention to these incidents was itself regrettable. We would have been better off, apparently, if each of the aggrieved parties had shrugged off the offense and deprived the public of further material on the subject. Furthermore, these critics argued, if offended minorities would turn the other cheek and the media would ignore such unfortunate but anomalous events their perpetrators would have no influence. According to this viewpoint, there is no *significant* current of tension between races in this country, only these flare-ups and their journalistic exploitation. As one columnist in the *New York Times* felt compelled to add, the Columbia professor might well feel violated by the placement of a noose on her doorknob, but she could hardly be justified in feeling "like a lonely voice in danger of being silenced" because "there's almost nothing at American universities more welcomed than a professor who specializes in the study of racism."[1]

Such observations run counter to almost everything I have gathered in studying the historic role of race in U.S. engineering since the mid-twentieth century. Race remains a major determinant of educational, occupational, and social experience in this country. Attainments by African Americans in the sciences are unquestionably dramatic and gratifyingly reported by academic, corporate, and media sources. Meanwhile, some universities' minority engineering programs are growing yearly. The long civil rights era, from roughly 1935 through Reagan's election in 1980, loosely conceived, did make considerable differences in both laws and customs around the country. But the representation of blacks among U.S. science and engineering graduates stayed low and remains low today; in some years since 1964 it has been static or even dropped. The number of black Ph.D.s in the science, technology, engineering, and mathematics (STEM) disciplines remains tiny.[2] *Diversity* has an extremely narrow meaning today in many science and engineering employment contexts, indicating an interest in ethnically diverse personnel but not necessarily in significant redistributions of economic opportunity or resources in or beyond the United States.

As for the uniquely supportive academic climate in which race scholarship supposedly now occurs, here too I am confused by the columnist's optimism. Certainly, the direct address of race in educational policy circles is often welcome today and race studies programs are now permanent parts of many U.S. universities, but these settings represent only a very small part of the academy. The choice to bring up matters of race in the laboratory or at the drafting table is puzzling to many people. To mention race in the university Office of Research, beyond discussions of initiatives targeted at minority inclusion, is anathema.[3] The absence of African American, Hispanic, Native American, and other minorities from many sites of science and engineering practice thus often exists without comment. This is not because the people who teach, patronize, produce, and deploy technological knowledge think ill of people of color and seek their exclusion from technical disciplines. Rather, it is because technology by and large remains a category of cultural activity that is supposedly raceless, undertaken in the classroom, laboratory, or workplace without regard to social matters such as practitioner identity or occupational equity. But engineering is constituted of a profoundly raced set of activities. In the nation's educational and employment sectors, decisions about technical talent and distributions of resources that determine individual opportunity follow lines of identity, providing more of those opportunities to some social groups than to others. The growing numbers of women and certain Asian minorities in American university engineering programs and technical jobs is sometimes invoked to challenge that socially inequitable characterization, but blackness is not commensurable with those identities. As this study will discuss, being black in the United States has historically carried intimations of weak intellect not associated with equally unfounded but influential stereotypes of so-called model minorities, such as persons of Jewish or Asian heritage.

Further, the entry of previously excluded groups into professions does not necessarily indicate the eradication of bias: when women and non-black minority practitioners are admitted to engineering schools, granted research funding, or hired for engineering jobs, they continue to face moments of discrimination not encountered by white, male engineers. Globalization heightens the impact of and drives these discriminatory ideologies, as technological employment and spreading market influence carry long-established U.S. majority interests to foreign settings in newly powerful ways. Yet to propose in a roomful of university administrators, engineering instructors, or even diversity-office personnel that some of the material accomplishments of this nation might be rooted in discriminatory agendas, both at home and abroad, draws quizzical looks. What can bigotry, or for that matter equity, possibly have to do with work at the laboratory bench, computer terminal, launch pad, or building site? For engineers and other scientists, these are places of profoundly practical cognitive labor where the very definition of rigor is felt to include the banishment of social concerns. The historian of engineering would ask, however, how such banishment could ever occur, given that we do not cease to be people of a particular gender, ethnicity, age, or income level when we arrive for work at the laboratory or launch pad. The cases in this book delineate this active role for identity in engineering education, following the ways in which conceptions of racial difference have configured admissions standards, course content, and definitions of rigorous teaching and research.

I became interested in the historical meaning of race in engineering because many of the most generous, liberal educators whom I have met (including some dedicated to minority inclusion) seem to consider notions of rigor and merit in academic engineering to be devoid of social character. Discrimination may well be happening on other levels, these colleagues suggest, causing doors to close well before minority students arrive at the university, but not in the actual assessment of students' technical talent or performance. Yet many educators in historically black colleges and universities have long understood the racial nature of performance metrics. As African American political scientist Charles Hamilton wrote in the *Negro Digest* in 1967: "And people will ask: What about accreditation? And I will answer: Our relevancy and legitimacy will accredit us. How creditable can the present system be when it cripples hundreds of thousands of black students annually?"[4]

I have not found much historical engagement with such concerns about the content of academic performance metrics in the years since Hamilton posed his questions, and I have tried to make them central to this narrative. Competence, skill, and identity remain fraught categories in many workplaces and communities, and are badly in need of historical study. Controversy surrounds the U.S. Supreme Court's 2009 decision confirming the validity of a civil service exam process in New Haven, Connecticut, that had placed only non-black firefighters in line for promotion. Some hailed the decision as a victory against "reverse discrimination," although others, including Justice Ruth Bader Ginsburg, found "substantial evidence of multiple flaws in the tests New Haven used." To consider the tests is to step back from reductive either/or logic that poses considerations of race as an alternative to considerations of skill. That kind of stepping back is very much the point of this book, and we can hope that in the coming months we may see a national climate warming to thoughtful debate on all such matters of social equity.[5]

THIS BOOK was written with funding from the National Science Foundation, with related projects receiving support along the way from the Society for the History of Technology and Drexel University. Among those who helped with the conceptual issues raised in this book were Bruce Seely, Janet Abbate, Scott Knowles, Greg Dreicer, Darin Hayton, Kali Gross, Rebecca Herzig, Jean Silver-Isenstadt, Agyeman Boateng, Antoinette Torres, Marissa Golden, Ray Fouche, Amilcar Shabazz, Thomas Zeller, Nina Lerman, and Jonson Miller. All of the scholars who attended a 2007 conference at Drexel on the historiography of race in science, technology, and medicine provided essential help, as have many of my graduate students. Conversations with Percy Pierre, George Callcott, and Elsa Barkley Brown steered the book's arguments in crucial ways. The single most important intellectual influence on this project was Elizabeth Hunt. At a formative moment, Lissa identified the study's core subject to be the historical linkage of standards of rigor and selectivity. In addition to being a beloved friend, Lissa was unquestionably the most incisive social analyst I have ever met and this book is inestimably weakened by her loss.

I must also thank the many archivists, especially Katherine Bruck at Illinois Institute of Technology and Anne Turkos at the University of Maryland, without whom vital bodies of evidence and interpretive frameworks would not have found their way into this book. Their creativity and knowledge assured whatever historical value it may have. I am also tremendously grateful to all the educators, administrators, diversity officers, university historians, and policy makers who agreed to be interviewed for this project. Michael Fisher at Harvard University Press and the manuscript's reviewers have contributed immeasurably to the book's readability and, above all, to any constructive potential it may possess.

Donald Stevens, Deborah Slaton, Harry Hunderman, Christine McDonald, JoAnn Stearns, Lynn Peterfreund, and the Rosenschein, Matsukawa, McKay, and Youngdahl families all lent invaluable support while the book was being written. I doubt if my daughter, Eleanor, even recalls a time when I was not working on this book; she has been exceedingly patient during its production. My husband, Peter Mayes, has been endlessly accommodating in logistical matters, but I am most deeply gratified by his support for the political agendas underlying the project. Finally, it is to the memory of my mother and father, Pearl and Leonard Slaton, to whom I dedicate this book. In addition to my mother's tireless editorial labors on the manuscript, she and my father supplied the motivations for this study: to ask questions about the political status quo and imagine a more equitable alternative. Their imprint is on every page.

List of Abbreviations

AAAS	American Academy for the Advancement of Science
ABET	Accreditation Board for Engineering and Technology
ACE	American Council on Education
ASEE	American Society for Engineering Education
AUF	Available university fund
BBHC	Benjamin Banneker Honors College (PVAMU)
CARR	Center for Applied Radiation Research (PVAMU)
CCUO	Chicago Committee on Urban Opportunity
EAP	Educational Assistance Program (UIC)
ECI	Engineering Concepts Institute (PVAMU)
ECPD	Engineers' Council for Professional Development
EEC	Engineering Education Coalition (NSF)
EIP	Early Identification Program (IIT)
HBCU	Historically black college or university
IBHE	Illinois Board of Higher Education
IIT	Illinois Institute of Technology
IITRI	Illinois Institute of Technology Research Institute
JSC	Johnson Space Center
LSAMP	Louis Stokes Alliance for Minority Participation (NSF)
MEP	Minority engineering program
MIT	Massachusetts Institute of Technology
NAACP	National Association for the Advancement of Colored People

NACME	National Action Council on Minorities in Engineering
NAE	National Academy of Engineering
NASA	National Aeronautics and Space Administration
NSF	National Science Foundation
NYA	National Youth Administration
OCR	U.S. Office of Civil Rights
PUF	Permanent university fund
PVAMU	Prairie View A&M University
STEM	Science, technology, engineering, and mathematics
TAMU	Texas A&M University
THECB	Texas Higher Education Coordinating Board
TSU	Texas Southern University
UIC	University of Illinois at Chicago
UMD	University of Maryland
UMDES	University of Maryland, Eastern Shore
URC	University Research Center (NASA)
UT	University of Texas

Race, Rigor, and Selectivity
in U.S. Engineering

CHAPTER ONE

Introduction

IN EARLY 2008, the National Action Council on Minorities in Engineering (NACME) issued its latest summary of data on minority representation in the nation's technical workforce. It is not, in tone or empirical findings, a particularly cheering report. The analysis, supported by a grant from the Motorola Foundation, documents downward trends or stagnation in the entry of African Americans, Latinos/as, and Native Americans into engineering degree programs and careers. These three groups have never approached parity with majority practitioners. Even at its peak, in 2000, African American representation in engineering careers reached only around 5.7 percent, while blacks made up some 13 percent of the U.S. population. According to NACME's recent findings, that participation has dropped to about 5.3 percent today.[1] Since its formation by a group of industry representatives in the early 1970s, NACME has tracked the educational and career attainments of women and minorities in a shifting national climate that at times has heralded civil rights or affirmative action reforms, and at times has withdrawn support for legal interventions based on gender or race. A summary of the data collected by NACME over the years indicates considerable positive change: for example, women have moved from comprising approximately 25 percent of engineering bachelor's recipients in 1966 to over 50 percent today. Between 1981 and 1999, all identified minority groups increased their representation in engineering. However, disaggregating that data reveals vast differentials by background. Freshmen undergraduate enrollments among Asian students increased at a rate of 137.2 percent in that period; among African American students, at a rate of 13.2 percent. As NACME president John Brooks Slaughter put it in his foreword to the 2008 report, there is still a "crying need for educational transformation in this country."[2] The founding goals of NACME, its corporate supporters, and many individuals in government and education committed to racial equity in engineering over the past four decades have yet to be fulfilled.

Some forty-five years after the Civil Rights Act of 1964, which was intended to eliminate racial differentials in education and employment, what are we to make of an occupational pattern that appears perpetually to follow lines of race? NACME's choice of title for its recent report, *Confronting the "New" American Dilemma*, both echoes Gunnar Myrdal's influential 1944 study of U.S. race relations and confirms that six decades later those relations remain troubled. Some areas of higher education and their related employment sectors, most notably education and business, have shown steady expansion in minority participation. But science, technology, engineering, and mathematics, or STEM fields, remain far less diverse than educators and policy makers might hope.[3] Efforts on the part of such concerned observers, ranging in scale from ad hoc mentoring of minority students to multi-university curricular initiatives and campaigns for increased state educational funding, have led, to quote Slaughter, to only "limited progress." Why have these efforts not made more of a difference? Furthermore, why have there not been more extensive attempts to bring minority citizens into STEM occupations? Statistical analyses such as those offered by NACME, the National Science Foundation (NSF), and many other organizations have offered important snapshots of educational policy and minority employment over the years, pinpointing sites of diversification or its absence. Individual educators and universities have also produced careful accounts of their efforts at minority recruitment and retention and associated classroom practices, while corporations and private foundations have chronicled their own interventions. If we are to understand the intractability of the dilemma to which NACME's analysts refer and achieve a sense of what has made some interventions succeed and others fail, a historical lens is needed to integrate these findings. As a step in this direction, we may consider the understandings of race among engineering educators and policy makers in the United States since the 1940s—a landscape of reformist ambitions playing out amid episodes of social stasis.

In this study I focus in particular on the opportunities encountered by African American aspirants to university engineering degrees. The intersections of black, Hispanic, Native American, Asian, and other ethnic minority

Introduction ⇁ 3

experiences are many, and of huge importance if we seek effective policy change. The meanings of age, gender, sexuality, national origin, disability, legal citizenship and residency status, and many other categories of practitioner identity have developed alongside and sometimes in tandem with these ethnic identities. Yet, I hope to illuminate certain cultural understandings of blackness in the world of majority intellectual endeavor and arrive at some sense of how desegregation, urban renewal, and affirmative action policies have expressed American ideas about blackness in particular.

For this reason, in this book I decline to discuss gender exclusion in any comprehensive way. Gender and ethnic minority status are often linked in discussions of America's "untapped science talent"; but, as noted, today women are entering engineering in increasing numbers, a pattern displaying some different social forces than those that surround the entry of blacks into the field. By making gender peripheral to this narrative, I show that blackness and whiteness in America are categorizations that persist in ways that other social differentiations may not. This is not a matter of ascribing relative amounts of influence to any of these differentiations (as if race has had a greater discriminatory role in U.S. education than has gender), but of probing the political instrumentality of each sort of differentiation. Nor is it to say that black and white identities are monolithic or representative of all racially informed experience in contemporary America. In fact, close study of one set of notions about identity in the United States (i.e., those surrounding the [in]compatibility of advanced knowledge work and blackness) may have a great deal to tell us about varieties of identity in other spheres, both domestic and global.

What History Offers

To understand how prevailing educational policies intersect with ideologies of race and social equity, this book is structured around a series of historical case studies. I consider a total of six engineering programs in three settings, beginning with the historically black and traditionally white campuses of the University of Maryland in the years immediately before and after the end of legal segregation. This is followed by a description of the engineering programs at the Illinois Institute of Technology and the University of Illinois' Chicago campus during that city's urban renewal efforts of the 1960s and 1970s. The closing cases are the engineering schools of the historically black and predominantly white campuses of Texas A&M University in the 1990s, when the first formal efforts to dismantle affirmative action policies emerged in that state. In each setting, I examine efforts to expand black opportunities in engineering teaching and research as well as obstacles to those reforms.

Obviously, these schools represent a very small sampling of the more than 2,600 accredited engineering degree programs in the United States today. Well over half of the engineering schools represented in that number offer some programming expressly for minority students. There is no question that a quantitative longitudinal study of these minority engineering programs (MEPs) could be designed to show which sorts of programs have yielded increased black participation in the sciences. Data amassed over the years by many public and private bodies, including NACME, the NSF, and the National Academies, would provide important contributions to efforts of this kind. But smaller ground-level qualitative studies such as this one can perform a different function: to reveal which features of higher engineering education figure in successful MEPs or prevent such successes. Those involved in the reform of racialized academic settings or in their maintenance over time have included university faculty, staff, administrators, and boards of trustees, as well as legislative, philanthropic, or corporate sponsors of higher education. Focusing on a few cases allows one to see how each group—sometimes in conflict with others, sometimes through consensus—contended with the issue of black underrepresentation in engineering.

Unlike policy studies and reports on individual minority educational opportunities or pedagogical best practices, this study highlights the institutional and cultural contexts in which educational change has occurred. It recounts moments in which daily educational operations confronted broad ideological issues, whether the educators involved saw minority inclusion as a vital social mandate or as an unfortunate distraction. In some instances, it is clear that educators' political leanings, whether progressive or conservative, inspired particular institutional or curricular policies. But the opposite causal pattern also exists. Admissions criteria, curricular priorities, physical plants, the patronage agendas of congressional funders or private donors, and reputational concerns are all highly variable among schools and over time, and all may have a bearing on how a given engineering school approaches the issue of racial diversity. Competition for students and research funding have mattered greatly in such decisions; thus, following changing ideas about those market conditions can help us capture educators' perceptions of how urgent, or not, the correction of minority underrepresentation in STEM fields may have been. These are all factors often missing from statistical assessments of minority higher education, which by definition must reduce complex social or historical situations to manageable metrics. Focus on six schools allows a close study of educators' day-to-day work while offering enough sites to provide meaningful comparisons.

Arrayed across several periods and regions, the specific universities or university systems help to demonstrate the multiple forces involved in Americans' understandings of race. We see, in high relief, the role of region (north or south), locale (urban or rural), patronage (public or private), and heritage (traditionally white or historically black) in university engineering programming. Pairing two schools in each historical instance controls for some factors (region or school system) while focusing on others (student demographics, institution patronage or prestige, etc.). In some instances, I will associate a particular policy choice made by one school with institutional conditions that many other engineering schools have faced. For example, a shift toward scientific coursework and away from practical or hands-on classes for students gradually pervaded engineering after World War II, largely dictating standards for rigorous curricula in U.S. universities. This costly and ambitious agenda for the discipline was disseminated through professional networks, and academic institutions were as often responding to as they were shaping these dictates.[4] By the 1950s, the Engineers' Council for Professional Development (ECPD; later the Accreditation Board for Engineering and Technology, or ABET) had centralized control over professional preparation, paralleling the exam and licensing procedures of other technical occupations such as architecture or medicine. The University of Illinois at Chicago tightened its admissions criteria after about 1970 to fit its sense of engineering accreditation structures and employer expectations. These are enduring and virtually universal conditions encountered by university engineering programs, and I know of no other literature in which educators' responses to these conditions over time have been discussed.

In other instances, I will show how a given school fits into its social or cultural milieu beyond the world of engineering or higher education. For example, in the early 1950s, University of Maryland President H. C. "Curly" Byrd collaborated with other segregated schools in the Deep South in an effort to establish a single, public black engineering school at Howard University. This attempt to fulfill the era's "separate but equal" legal requirements for black education highlights some of the pressures perceived by many southern conservatives. By following conceptions of racial equity through time in a single educational field, we may gain a sense of where social ideologies yielded to existing institutional conditions and, alternatively, where universities and their faculties or sponsors enacted those ideologies.

Justifying Diversity

This book does not recount the experiences of minority students in their own words or with the same degree of detail that it brings to educators' experiences. A small number of educational specialists have studied the ways in which minority students make choices about their career paths, and rather than add directly to that data I offer instead a historical backdrop for further such study.[5] In a free society, each young person chooses his or her occupation on the basis of multiple factors, including family tradition, intellectual interests, available resources, and access to training and employment. Individual choice has intersected with societal conditions as college-bound African Americans in every generation have tended, more than white students, to pursue occupations other than engineering. As civil rights protections found a foothold and discriminatory college admissions and hiring decreased after 1954, blacks strengthened their representation in degree programs that led to educational, civil service, and nursing careers. The overall number of college degrees granted to African Americans has gone up steadily since the 1960s; most recently, there have been considerable gains in business degrees granted to blacks.[6] A historical perspective may lead to an understanding of why engineering has remained a less attractive option to black Americans seeking college degrees than to other groups. Engineering has long offered Americans a professional path toward greater economic opportunity; from its origins as a degree-granting field in the late nineteenth century, engineering saw countless sons of rural, and then immigrant, Catholic and Jewish families transition from backgrounds in agricultural or wage labor to salaried employment. In addition, after 1970 or so, women, Asian, and South Asian students saw significant gains in engineering, and these increases continue today. That African Americans, Hispanic Americans, and Native Americans have not followed that pattern suggests that ethnic identity remains a factor in STEM participation.[7] This book offers a picture of change over time to help delineate those mechanisms of difference, on the presumption that racial and ethnic diversity are desirable goals for the nation's STEM labor force.

That presumption itself has a history that has been little studied but which reflects changing social relations in the United States. Over the past fifty years, three types of justifications have commonly appeared in defenses of racial diversity in technical fields: legal, economic, and social justice arguments. Invoked by educators, employers, and policy makers, each sort of justification has waxed and waned in popularity through wartime, economic crises, and widely varying levels of national concern regarding civil rights. Although I will consider some of the legislative fallout from the past several decades, readers will find relatively little analysis of legal arguments in this book. State and federal policies governing the provision of higher education to minority citizens have included desegregation, urban renewal, and affirmative action policies, and innumerable legal contests have pitted educational institutions against excluded students. Since the 1978 *Bakke* case, suits have arisen from aggrieved white applicants as well as from excluded minorities.[8] But the regulatory climate in which U.S. universities have operated since 1940 is a second-order phenomenon that follows from sociopolitical trends on city, state, and national levels. Economic and social justice arguments for racial inclusion, in contrast, express sociopolitical priorities more directly, and I address those arguments explicitly.

Economic arguments for greater black participation in engineering, based on the need for enlarged technical labor pools, will likely be compelling for many readers today. Through World War II, the Sputnik era, and subsequent seismic shifts in distributions of global economic advantage, proponents of greater black inclusion in technical occupations have cited the importance of maximizing the size of the U.S. technical labor pool. In this view, the nation requires the greatest possible access to talented practitioners in order to attain its greatest possible economic competitiveness. American corporations have overwhelmingly given this reason for their vocal support of diversity efforts in education and hiring. Anxieties about national security have been closely aligned with those regarding technical innovation and productivity. As was the case during World War II and at the height of the Cold War, these concerns have taken on particular urgency for their supporters since September 11, 2001.[9] It is on the basis of these various labor force arguments that a majority of MEPs (and programs for women in engineering) have been founded and funded since the 1960s. The social justice rationale for such interventions has been a less consistently persuasive one. In this sort of argument, Americans must eradicate all traces of racial injustice, including minority underrepresentation in science and engineering arenas, for moral rather than strictly practical reasons. Provisions for educational reform based on social justice arguments do not aim for economic return at the corporate or national level. Productivity gains may well follow from race reforms, but these gains are of secondary importance to matters of individual equity.

As the recently coined term *post-racial America* implies, such concerns have less of a grip on many Americans today than in previous decades. In the 1990s, reformers faced growing opposition to affirmative action, minority set-asides, and other race-based interventions in education and hiring, with some white citizens and politicians even claiming "reverse discrimination." The media may have given disproportionate attention to those raised voices, but certainly unalloyed ethical or justice arguments in favor of racial inclusion would have had much greater resonance a generation or two ago during the peak years of civil rights activism. From the normative standpoint of this study, it is this relative weakness of current social justice ideologies in certain quarters that demands historical explanation; such explanations may give this book relevance beyond engineering. Social justice goals, considered alongside economic imperatives, could support dramatically expanded educational and employment opportunities for minority citizens. In turn, if engineering becomes a profession with a pronounced culture of social justice among its practitioners, it would seem likely that benefits for other members of society will accrue. Engineering enacts and shapes the nation's infrastructural, industrial, and military agendas. If technical experts come to their work with an enhanced sensitivity to fairness and democracy those realms may show an increasingly equitable character. The causes of class, gender, ethnic, and global equity will be well served if influential experts make fairness a priority alongside the traditional mandates of efficiency and productivity.

Economic analysts may conventionally pose social change at best as a second priority after profit and efficiency, and at worst as antithetical to those goals (traditions in corporate philanthropy notwithstanding). Radical historians go so far as to say that corporate culture, or capitalism itself, may simply be incompatible with democratic values. But I am not proposing that only noncorporate, noncapitalist economies can support authentic democratic reform. Instead, I imagine a more equitable profile both for technical disciplines and for the corporate or state patronage networks in which they operate. Progressive change in each realm will support change in the other.

In projecting constructive changes in U.S. race relations through changes to engineering institutions I hope not only to challenge those with radical antitechnology social agendas, but those—on the other end of the political spectrum—who see the country today as well beyond the racial problems of earlier eras. For certain observers, concerns with social justice are inseparable from radical "redistributional dreams" that undercut the essential features of science, technological progress, and industrial capitalism; these are among the core arguments of so-called culture warriors.[10] Others in the science and technology establishment feel that conditions for blacks in higher technical education remain discriminatory. Among these voices are those of NACME president Slaughter, Freeman Hrabowski, and Shirley Malcom, all influential figures of long experience in higher education and technical research arenas. Beyond engineering and education, we are hearing similar concerns at this writing from many supporters of Barack Obama (not to mention from those activists who find his policies to be too conciliatory). Racial equity is not off the table in U.S. public discourse, but it faces formidable countertrends; its place in the public consciousness is in flux. For this reason, this book not only recounts the histories of efforts to diversify engineering occupations, but also follows the histories of different justifications for those efforts.

Critiquing "Eligibility"

Social justice rationales for educational reform not only arise from broader political ideologies than do workforce rationales, they may also lead to a more profound understanding of race-based inequities in STEM fields. On some level, calls for greater minority presence in STEM fields on the basis of America's workforce needs offer unassailable reasoning. The exigencies of international business competition or global security require a greater pool of technical workers than is currently available. The shift of many regions around the country to "minority majority" status exacerbates the problem; without intensive recruitment of previously excluded minority groups, parts of the nation will find themselves without sufficient labor pools. In periods (such as the 1960s, or the early 2000s) during which young people trend away from choosing engineering as a career, further pressures arise to bring new populations into STEM occupations. However, supply and demand analyses of the American technical workforce by and large fail to critique what counts, in any given time and place, as eligibility for STEM participation. Those notions of eligibility—which explicate historical ideas of intellectual value, or individual merit—are central to this book.

Slaughter has indicated that exploration of what constitutes legitimate knowledge and skill in a given setting will reveal a central instrument of minority exclusion from STEM occupations. According to Slaughter, the current crisis is one with multiple causes. Today, fewer young Americans of all ethnic backgrounds are choosing engineering careers. However, shrinking federal, state, and institutional commitments to financial aid for disadvantaged students, increasingly competitive college admissions, and a lack of access to sufficient high school preparation in math and science among minority communities over the past decade all contribute to disparities in numbers of black, Hispanic, and Native American engineering graduates. He adds that recent attacks on the use of race in academic admissions, scholarships, and programming have exacerbated the situation. Although Slaughter points to the necessity of changing government and university policies, he is also clear that the *content* of what we consider to be optimized engineering teaching and practice may also be implicated in minority exclusion. Those valuations about what should and should not be taught in the engineering classroom or laboratory have a great deal to do with who has counted in American colleges and universities as a well-qualified candidate for engineering degree programs. As Slaughter puts it, our understanding of merit has been associated closely with privilege and has produced a "narrowed" notion of what constitutes excellence.[11]

As Christopher Newfield has shown to be the case in many academic arenas, the invocation of customary ideas of merit or performance standards encourages the maintenance of existing opportunity structures in STEM fields. Those standards constrain minority opportunities in two ways. First, as Slaughter points out, they narrowly define merit in a way that excludes many young people who come from disadvantaged backgrounds. Second, such standards arbitrarily link university selection processes to rigorous technical practice, ignoring choices being made about curricular content. Historically, engineering educators have rejected remedial instruction, widened admissions criteria, and collaborations among schools on the basis of such standards. Those choices lead to demographic narrowness. At the same time, faculty members have commonly found that mentoring or advocating for minority students jeopardizes their research or limits their credibility within academia. Significantly, and in contrast to many other studies of minority experiences in higher education, I focus not only on institutional change (and stasis) over time but on lending analytic equivalence to teaching and research activities in the university. In this way, I hope to reveal multiple characteristics of scientific and technical thinking in modern U.S. culture that devolve onto matters of identity. These characteristics are displayed in both engineering pedagogy and research, and crucially, in the relative status each holds within a given university or accreditation structure. This dual focus ensures that we do not consider such categories as enrollment figures, graduation rates, educational "outcomes," or students' "life chances"—all commonly used to assess the success of minority engineering programs—as somehow independent from the content of engineering knowledge itself.[12]

That attention to content may show the historical process by which rigor and selectivity have come to be so closely associated with each other in American engineering.[13] Suggesting that standards of technical talent have social bases and impacts may seem counterintuitive. After all, it is hard to imagine that anyone today would say that a bridge stands or falls based on the ethnicity of its designer. We would far more readily say, if we were to consider the matter at all, that in order to stand the structure's stresses and strains must be correctly computed and its materials and construction closely monitored. If design or calculation or construction errors occur, the bridge may fall; if not, it will stand. Yet engineering is nonetheless an intellectual realm in which practitioner identity plays a role; within the admissions criteria, curricular emphases, and classroom exercises of most engineering programs, something is working against proportionate black involvement. Metrics of eligibility and other content-dependent matters in engineering require examination if we wish to find out just what that might be. Pipeline explanations, which attribute low minority participation in engineering to deficits in primary and secondary schooling, tell part of the story but fail to account for activities at the postsecondary and professional level. We know from previous quantitative research such as NACME's that some college-level interventions do bring more minority students into engineering disciplines. What we do not yet know, from the perspective of changing race relations in the United States, is why there are not more or larger such programs, and thus more engineers of color. The six historical cases that make up this study begin the work of answering that question.[14]

Case Studies

In this study of engineering education and widely held racial values in the United States, I leave the stories of many commendable but exceptional individual institutions to others. For example, Howard University, in many generations the nation's leading producer of black engineering degree holders, did not generally reflect the conditions under which most engineering schools at historically black colleges or universities (HBCUs) functioned. Some of the schools discussed here, which I judge to be more typical of HBCUs that have aspired to technical excellence, may be unfamiliar to readers. At the same time, many of the educators described in the following chapters, both conservative and progressive, were people of considerable influence in the world of higher engineering education. Their responses to prevailing conditions of college-level technical education reached wide audiences and had significant ripple effects in pedagogy and policy circles. Individuals, institutions, and programs that have been marginal contributors to mainstream engineering pedagogy as well as those that have been central all have something to add to our understanding of racial equity, and examples of each appear among the cases which follow.

The narrative opens with a study of the University of Maryland (UMD) system as it operated through the 1930s, 1940s, and early 1950s. In Chapters 2 and 3 I track differentials in facilities for black and white students in this land-grant university system. As a border state, Maryland was a locale in which integrationist voices grew louder through the mid-twentieth century, but simultaneously faced sturdy opposition. The National Association for the Advancement of Colored People (NAACP) was very active in Baltimore, while UMD system president Curly Byrd found vigorous support for continued segregation at the system's main campus at College Park (where, he warned ominously, Negro enrollees would come into contact with "500 [white] girls"[15]), close to Washington, D.C. In addition, while administrators modernized technical instruction at College Park during the 1930s and early 1940s, the system's black campus, in the town of Princess Anne on Maryland's Eastern Shore, remained deeply disadvantaged in all fields. By associating the state's minority citizens with agricultural and low-level industrial occupations, Byrd, UMD's Board of Trustees, and Maryland's legislature justified the wholesale denial of state resources to the black land-grant campus. Byrd's increasingly desperate attempts during and after World War II to preserve the old social order in light of encroaching civil rights activism and legal reforms that challenged racial segregation are discussed extensively in Chapter 3. The UMD case illustrates the essential compatibility of programs for economic modernization in the south and ongoing white supremacist inclinations during these years.

Before mounting legal pressures forced him to integrate undergraduate engineering programs at College Park, just prior to *Brown v. Board of Education of Topeka, 347 U.S. 483,* Byrd engaged in a bizarre effort to promote the state's economic development as a project disunified by race. The implications of that disunity for the presumed universality of technical and scientific and intellectual enterprises are striking. Such logic revealed the depth and epistemic impacts of racist ideologies in the segregated university system. While the retrograde elements of Maryland society and government actively promoted racist ideologies in operating the state's land-grant schools, some beliefs about black ineligibility for scientific careers were deeply held by integrationists as well. Through the 1940s the federal government, particularly through the U.S. Office of Education, supported the creation of new opportunities for minority Americans in higher education. Yet these same policy makers expressed serious doubts about the ease with which black citizens, long deprived of equal educational opportunity, might "adjust" to the demands and mores of research professions. Race retained its significance as an index of intellectual potential. The allure of scientific racism, if now diluted by the rejection of certain biological theories, had not entirely disappeared (as remains the case today). William H. Watkins's analysis of black education conducted under white direction reminds us that this too was an example of how the ostensible objectivity of scientific research was used to consolidate white privilege.[16] It is necessary to consider how some self-identified progressive interests helped maintain discriminatory ideas about minority qualifications for scientific achievement after World War II in preparation for the study of the "true" civil rights era of the early 1960s through the late 1970s. The ongoing ambivalence of programs for black inclusion in U.S. engineering education through that activist period forms the basis of the next case under consideration here.

During the 1960s and 1970s, unprecedented civil rights activism, profound doubts about U.S. involvement in Vietnam, and other challenges to the emergent military-industrial complex found especially strong expression in U.S. cities. Together, these affronts to long-held national cultural priorities implied a redistribution of opportunities and resources in America along more equitable lines than had ever been the case. Women, Native American peoples, farmworkers, and gay and lesbian groups all built social movements on this growing energy. Some centuries-old class inequities seemed about to give way, and education, particularly urban education, was imagined by many to be a central instrument of such change. Engineering programs that took shape in two Chicago schools in this era are the focus of Chapters 4 and 5: the University of Illinois' new "Chicago Circle" campus (UIC; the nickname referred to the nearby "Circle" highway interchange), opened in 1965; and the older, private Illinois Institute of Technology (IIT), which began innovative programs for minority engineering students in this period. Differences between ideologies at UIC, initially operated along very inclusive lines intended to correct long-standing educational imbalances in Chicago, and IIT, a school that sought heightened status as a site for elite research in this era, are telling. Similarities, however, which became evident as the schools' agendas converged through the 1970s, are also significant. Together their programs point to prevailing ideas of what constituted quality engineering in this era and the enduring importance of practitioner identity in such judgments about eligibility or achievement. Ultimately, as Chapter 4 recounts, neither engineering school transcended the hierarchical occupational forces that had historically privileged white, male practitioners in engineering, despite the progressive intentions of many of their faculty members and administrators.

Reformist approaches to engineering pedagogy at UIC and IIT broke with convention by legitimating considerations of race in college admissions and academic services. Yet neither institution engaged with the day's most transformative notions of identity politics, which saw content regarding black heritage as a desirable element of all black education. In Chapter 5 I describe how expressed values about what constituted appropriate disciplinary boundaries and engineering research subjects at the two schools helped keep racial issues out of the spotlight. During this period an old firewall between humanistic and technical work in Western culture found new supporters. Profound concerns among some UIC and IIT faculty about the detrimental social impacts of science and technology, and about the social makeup of technical occupations, were countered by deeply ingrained and linked conceptual and institutional habits in engineering.

If Chicago's universities could not enact lasting changes to the racial profile of U.S. engineering through the activist 1960s and 1970s, it is perhaps not surprising that the 1980s and 1990s saw even fewer attempts to subvert old national patterns of minority underrepresentation in technical disciplines. Those decades saw a reassertion of dominant ideas about academic eligibility and attainment that took the form of vigorous objections to and legal reversals of affirmative action as conservative political voices gained new credence around the country. From Reagan through the two Bush presidencies, and certainly with some support from Democratic administrations in between, reformist political agendas lost much of the support they had lately accumulated among U.S. politicians. Voters either inspired or seemed willing to follow these conservative paths, finding comfort perhaps in their nationalist overtones as the fall of the Soviet Union destabilized foreign relations. September 11, 2001 no doubt heightened the appeal of nationalist ideologies for many. But the nation also experienced a conservative retrenchment on domestic issues after 1980, signaled by increasing distaste for social welfare programs and, among many majority citizens, for race-based affirmative action programs in education and employment. What caused this retrenchment? World oil prices and the entire U.S. industrial outlook fluctuated in these decades, but economic downturns in other eras, such as the 1930s, had produced just the opposite impulses in many voters. Some deeper strain of fear and isolationism seemed to be reasserting itself among majority Americans as the twentieth century drew to a close. Paired with that discomfort was a devaluation of "politically correct" multiculturalist sensibilities that were now seen by some as constraining free speech and civil liberties, or as fragmenting a supposedly once-unified culture.[17]

As we reach the end of the first decade of the twenty-first century, the revelations of government neglect of minority communities made so potently by Hurricane Katrina seem to have had few lasting impacts on this quiescent atmosphere. The proliferation of charter schools attests to lowered public expectations for governmental provision of equitable educational systems. In the final case study, I describe race relations in the Texas A&M University system through the eighties and nineties, tracking the return (or continuance) of traditional ideas regarding minority absences from intellectual fields.

In comparing the main campus of Texas A&M University at College Station (TAMU) and the system's historically black branch at Prairie View (PVAMU), the reader will see some familiar patterns. As Maryland's land-grant system had in the early 1950s, the Texas system in the 1990s displayed fidelity to the state's racialist history, even as the twentieth century closed. As recounted in Chapter 6, TAMU president Robert Gates (currently the U.S. Secretary of Defense) spoke of welcoming diverse populations to higher education in Texas, but he put "standards" front and center. He framed this as a matter of promoting the "color blindness" that many now saw as the road toward improved race relations in the country.[18] Merit in this formulation would be the way out of race consciousness, seen here as a crippling vestige of a bygone era. In making this claim, Gates stigmatized discussion of inequity or discontent that might point to systemic problems within the Texas A & M schools or U.S. higher education as a whole. Scholarships for disadvantaged TAMU students and minority recruiting programs grew under Gates' leadership, but where race blindness is a stated ideal, to speak of race is to bring on oneself suspicion of racist intent. It is hard to imagine a more effective means of suppressing inquiry into the role of identity.

But, as is argued in Chapter 7, conventional metrics of engineering preparedness and merit were not going unquestioned, even in this politically conservative setting. Some innovative, NSF-sponsored programs for minority engineering students at TAMU survived as faculty persisted in these efforts despite institutional disincentives. Engineering departments at PVAMU and programs for the support of research at HBCUs by the National Aeronautics and Space Administration (NASA) joined in a strikingly progressive project. PVAMU in the 1980s and 1990s continued to celebrate the sense of black heritage that had characterized it through previous generations. An underresourced campus for much of its 100-year history, much like Maryland's Eastern Shore campus, PVAMU in the 1990s built on much-delayed state appropriations and created centers of engineering research of national renown. NASA supported this unusual amalgam of cultural and intellectual programming with a notably flexible set of standards for grant support to HBCU science and engineering research. The result was a rapid buildup of competitive research work at PVAMU and other traditionally minority institutions. How and why NASA undertook this relatively radical approach to enhancing minority participation in high-tech research, and the ideological challenges PVAMU thus posed to white-dominated engineering in this era, are the subjects of Chapter 7.

As all these cases show, lines between segregated and desegregated settings, between south and north, between pre- and post-*Brown* eras, between conservative and progressive ideologies, are permeable. Some of the priorities found in baldly discriminatory racial policies were not entirely distinct from those featured in ostensibly progressive doctrine. Some of the bases of black exclusion from engineering at work in the 1930s may still be seen at work in universities and workplaces today. Dismayingly, educational settings in which egalitarian reforms might have been expected to find fertile ground have displayed distinctly retrograde inclinations, leaving interventions on the behalf of minority citizens weakened or absent altogether. This is the case even in the early 2000s, both hailed and derided as the "post-civil rights era." Radical reformers and black separatists since the 1960s have perhaps been justified in turning away from established sites of higher education altogether, seeing U.S. colleges and universities as too deeply embedded in systems of white privilege. Surely the lack of educational and occupational choice with which many African Americans have lived in these decades is reprehensible, and all authentic efforts to correct that lack are laudable. But that does not mean that conventional practices in higher education have ever held the potential of profound reform, as some critics of U.S. academic and scientific institutions have suggested. The case of PVAMU and NASA's partnership points to interstices in this system in which liberal reforms might take root. However, what does it mean that this departure from conventional standards of merit in engineering played out under NASA's auspices? Do the engineering projects associated with the space program, in fact, perpetuate innately oppressive social agendas? After all, these projects support the industrial and defense establishments first and foremost, even if they yield some benefits for larger civil society in the form of, say, new medical technologies. With a number of African American, Hispanic, or Native American persons entering engineering occupations under NASA's programs (and at this writing, General Charles Bolden, an African American, nominated by President Obama to be head of the agency), exactly how likely are larger and lasting assaults on the nation's discriminatory social structures to follow?

We thus return to some of the most difficult questions associated with issues of racial representation in engineering, and articulating these is my goal in Chapter 8. What exactly is the relationship between practitioner identity and practice in engineering? That question may be added to other historical inquiries that ask *how* strong, or *how* determined, the relationship between technological knowledge and the maintenance of a discriminatory society may be in a given time and place. Proponents of diversity in engineering have claimed for the past thirty years that it is not only more talent we need to solve the nation's domestic and global challenges, but talent that represents a wider range of interests than has been customary. We are really asking what is meant by "wider" here. In his study of boundary maintenance among scientific fields, Thomas Gieryn aptly depicts demarcation as a defensive strategy. For example, scientists preserve their decision-making authority when they declare lay persons or nonscientist experts to be differently, and potentially inappropriately, equipped to make certain kinds of decisions. As superbly suited for professional gatekeeping as these demarcation activities may be, however, for some scientists, engineers, funders, and politicians such strategies serve an *offensive* purpose. Such acts, discrediting the knowledge or ideas of certain groups, may position socially marginal citizens in economically marginal occupations, and (in the United States, at least) function as part of the ongoing cultural process of race formation. That process, as described by Michael Omi and Howard Winant, is proactive and involves operations at the institutional level. In both respects, the racialized worldview is emphatically not the product of individual pathologies. Historians and sociologists of science have identified patterns of stratification in the scientific laboratory that constructed professional boundaries as well as structural inequities of value to elites across many occupations. This model has yet to be applied to engineering classrooms or laboratories.[19]

It takes very little exposure to modern media, corporate public relations, and educational discourse to know that usage of the term *diversity* has vastly increased in the past few years. In some settings, it stands for profound challenges to long-standing social structures: the goal of recognizing and reforming some familiar racially privileged avenues to intellectual and economic attainment. In other settings, it suppresses conflict with an amazing thoroughness. Conceptualized as one route toward improved innovation, economic competitiveness, and a more reliable pool of scientific labor, diversity programs frequently celebrate superficial differences in experience or cultural heritage. Even when well intentioned, such celebrations may foreclose inquiry into the racialized functions of conventional gatekeeping tools such as standards of talent and eligibility. Such diversity projects thus reify the very social structures that have closed technical professions to minorities for generations. Recognizing the term's multiple political utilities may launch readers on a set of still further questions about the powerful role of race in U.S. engineering.[20]

James D. Anderson's seminal study of black higher education before 1935 offers a valuable starting point for a historically informed approach to these questions. Regarding the extensive educational opportunities available to most white Americans and the severely discrepant schooling systems provided for black citizens since the nation's inception, he notes: "These opposing traditions were not, as some would explain, the difference between the mainstream of American education and some aberrations or isolated alternative. Rather, both were fundamental American conceptions of society and progress."[21] In other words, in a circular sort of logic, white educators through the 1930s held that where blacks were taught, intellectual achievement could not be found; where such achievement was desired, blacks must not be taught. Standards for rigorous teaching and learning helped naturalize black absence from science and technological enterprise, as from many other sectors. This historical narrative will help reveal whether and how conditions in higher technical education have changed in the decades since.

CHAPTER TWO

Identity and Uplift

*Engineering in the University of Maryland System
in the Era of Segregation*

From the 1930s through the 1950s, Americans displayed a profound ambivalence about the correction of race-based educational inequities. During those years federal policy makers bemoaned the shrunken and underfunded facilities provided for African American education, but they offered no clear indictment of segregation as a wellspring of such inequities. The nation fought a world war predicated on the idea that the United States would be a global "torchbearer of democracy," but then sustained Jim Crow laws for publicly funded education in parts of the country for a full decade after the defeat of fascism in Europe. And somehow, year after year, it still seemed possible to many educators, lawmakers, and voters in both the south and the north that racially separate educational provisions for black citizens could also be equal to those supplied to white students, despite all evidence to the contrary. At the very least, that fiction was tolerable to many, including a federal government unwilling to alienate its southern constituencies. Among the general white public, only committed civil rights activists posed strong objections to that bifurcated system.[1]

By the 1950s, higher education was understood to serve a powerful economic function in the nation, expanding from an elite purview to a site of opportunity for working-class Americans. In the science, engineering, and medical fields, the societal benefits of higher education were particularly clear to many observers; however, its limited availability to minority youth contradicted ideologies of both personal uplift and national development.

On close examination of the engineering education at the University of Maryland, a public land-grant university with a heritage of serving the state's economic needs, one can see the instrumental role that higher education held in both the state's economy and its stratified social vision. Technical teaching and research in the UMD system, as in U.S. land-grant schools in general, represented a means of accumulating new knowledge and a skilled labor force. But in a climate still tolerant of segregation, Maryland's university system also offered the state's white leadership a means of assuring a racially stratified workforce.

As with many U.S. universities of this era, the UMD system, with its origins in the Morrill land-grant legislation of 1862, was a site of steady growth in the fields of science and technology after 1920. From its early emphasis on agriculture, the UMD system expanded its involvement with teaching and research to serve the needs of government and industry. The institution helped to diversify the state's economy and cultivated interest from the nearby federal government and national trade groups. The system's main branch in College Park, very near Washington, D.C., grew after World War II with the dual intentions of democratizing higher education and ratcheting up standards for employment in many sectors of the economy.[2] UMD's professional schools of law, medicine, and nursing, all located in Baltimore and with origins in the first half of the nineteenth century, also grew in this period. In this development the UMD system resembled many public universities that enlarged with help from the federal GI Bill and a burgeoning interest in the expansion of U.S. scientific and technological capacities through higher education. But Maryland is a border state; in many respects its southern social affinities persisted to create a distinctly inegalitarian institution. No agenda of expanded opportunity and technological development characterized the UMD system's campus for Maryland's black students, located in the town of Princess Anne on the state's Eastern Shore. The system's main campus was closed to nonwhite students from its inception in the 1860s through the 1950s, as were its professional degree programs in Baltimore through most of this period.

The notable exception to that pattern was the admission of black students to the UMD School of Law following a U.S. Supreme Court order in 1936, the result of the NAACP-led case of *University v. Murray, 169 Md. 478*. That no other effort at integration followed this startling development in the UMD system for the next fifteen years has much to tell us about the power of discriminatory beliefs in the university in this era. With that one exception, the state's black citizens could attend only the "colored" Eastern Shore branch of the UMD system. The Princess Anne campus was designated as part of the state system under the second Morrill Act of 1890 for the establishment of land-grant colleges for Negro youth. According to that legislation, land-grant universities closed to black citizens had to maintain such branches in order to receive federal funding. With the passage of the act, all seventeen southern and border states that had legally mandated dual education systems set up land-grant colleges for Negroes. Publicly supported alternatives for black college students in Maryland before the 1954 *Brown* decision also included Morgan College in Baltimore (later, Morgan State College) and two black state teachers' colleges. The histories of these three schools are closely intertwined with that of the University of Maryland Eastern Shore, as that campus is known today (it was known first as Princess Anne Academy, and for some years as Maryland State College; it became a formal branch of the UMD system in 1948). Despite Baltimore's development as a site of significant civil rights activity in the 1930s, particularly through the work of the NAACP and the young Thurgood Marshall, the UMD system managed to maintain its influence among conservative Maryland legislators and other supporters to sustain a set of racially segregated institutions of drastically uneven quality.[3]

The mid-century development of science and engineering at UMD displays a striking combination of modernizing and retrograde ideologies. Between 1935 and 1954, the entire UMD system was under the painstaking direction of President Harry Clifton Byrd. Known as "Curly," Byrd was famously affable but also a notorious micromanager. Wedded to traditional southern social values, he fought for segregation in Maryland's land-grant university system as state and federal courts increasingly ruled against dual education provisions. Under his presidency, the system's Eastern Shore campus (hereafter, for simplicity, Eastern Shore) remained largely without resources until pressure mounted for separate but equal black educational facilities in the late 1940s. Throughout his tenure, as Byrd assiduously developed engineering at the all-white College Park campus of the UMD system (hereafter, College Park) and new relationships with private industry, Eastern Shore remained almost entirely agricultural and trade-oriented in its curriculum. In those curricular differences lay the kernel of Byrd's justifications for a two-tiered public university system: for the UMD system's president and its board of trustees, and for the state legislators who stood behind them, blacks would fill one set of occupational roles, whites another. That the former jobs were almost universally lower than the latter in pay, prestige, and intellectual challenge both justified and generated discriminatory racial ideologies.

That scheme, which linked different kinds of gainful employment and intellectual activity to different races, echoes in several ways the operations of many HBCUs, both public and private, in the United States from the Civil War onward. These schools, often run, in William Watkins's term, by "the white architects of black education," commonly followed the discriminatory ideologies of white industrialists and politicians hoping to confine black citizens to the lowest echelons of economic participation. Although the decades between 1930 and 1960 were unquestionably a transitional period in which civil rights found a purchase in many communities, north and south, Byrd's career at the University of Maryland demonstrates that U.S. efforts to provide minority higher education even at the end of that period consolidated white civic and economic power through the control of knowledge and occupational opportunity.

Engineering is not the only field in which blacks experienced major disparities in educational opportunities in this period. As Oscar Chapman concluded in his extensive 1940 doctoral dissertation appraising segregated land-grant education in the seventeen southern states, "There is a total absence of the following types of work for Negroes, which are included in the educational offerings of the land-grant colleges for white students: (1) architecture; (2) dentistry; (3) engineering courses; (4) forestry; (5) journalism; (6) law; (7) library science; (8) medicine; (9) pharmacy; (10) veterinary medicine; (11) nursing; and (12) commerce and business."[4] Nevertheless, the focus on engineering provides a particularly clear picture of how policy makers and educators have understood connections among students' intellectual capacities, their occupational potential, and their race and gender. By the 1930s, scientific racism per se, which explained blacks' lowered economic status in the United States on the basis of their supposed intellectual inferiority, was passing from favor. Nonetheless, with countless declarations on the condition of black higher education and levels of black economic attainment, white educators and policy makers continued to construct African Americans as intellectually and behaviorally ill-suited for advanced technical labor.[5]

In this chapter, I examine the institutional choices that assured the absence of black students from College Park through the years just prior to World War II. In the eyes of people like Byrd, who saw segregation as the best option for Maryland, economic and social planning dovetailed, casting higher education for the state's minority citizens as a low priority. Engineering emerged in the UMD system as an area reserved for whites. The pressures that the war brought to bear on segregated higher education are discussed in the following chapter, as well as the limits of the war's impacts as separate but equal facilities for public higher education gained credence in Maryland.[6]

Origins of Land-Grant Education on the Eastern Shore

In 1886, Methodists of the Centenary Bible Institute of Baltimore founded a school for "men and women of African descent" in the town of Princess Anne, to be operated by Morgan College, a private black college located in Baltimore. The Delaware Conference Academy, as the school was first called, occupied a small plot of land at one end of the town and became known as the "Industrial Branch" of Morgan College. According to Carl Person's history of higher education at Princess Anne, the school underwent six name changes between 1886 and 1970, "with almost every change signaling a derailment of previously defined and cultivated course or a question of the institution's purpose."[7]

In 1890, the Second Morrill Act required that any state receiving federal funding for a land-grant university closed to nonwhite students create a second facility for its minority citizens. The Maryland state legislature, eager to retain federal support for the all-white Maryland Agricultural College located at College Park, complied by designating the school at Princess Anne as Maryland's black land-grant school, keeping it under the administrative direction of Morgan College (rather than making it a subsidiary branch of the school at College Park). Now receiving one-fifth of the federal funding provided to the state of Maryland for its two land-grant schools, the Eastern Shore campus began to offer courses of study in home economics and industrial and mechanical education, producing teachers in these areas that met with the approval of the Maryland State Department of Education. Those teachers, of course, were intended for employment in the state's all-black primary and secondary schools. By the end of the decade, the campus had nearly 100 students receiving instruction from six faculty members in education, agricultural subjects, and trades such as shoemaking, carpentry, cooking, tailoring, and blacksmithing. A small liberal arts curriculum accompanied these practical courses of study, but as was the case at most public and many private institutions for black youth that were directed by whites, skills that might lead to immediate employment were favored over traditional humanities or classical subjects by those overseeing course content.[8]

In a number of ways the school followed what many scholars have seen as the prevailing ideal of industrial education for black youth around 1900, an ideal often associated with Booker T. Washington's instructional aims at Hampton Institute or the Tuskegee Institute. This pedagogical approach stressed training in areas in which black Americans were most likely to find employment: the trades, domestic service, and teaching. It is a vision often contrasted with that of W. E. B. Du Bois, an advocate for more ambitious professional education and immersion in a far wider range of cultural subjects for African Americans.

The practical emphasis at Eastern Shore could also be associated with retrograde beliefs about the uneducable nature of African Americans, ideas supported by many white southern policy makers well into the twentieth century. What is more, prior to about 1920, the majority of instruction offered in the Negro land-grant schools, including Eastern Shore, was at the high school level, further reducing the probability that black students would attain public university educations comparable to those offered to white students. But historians have recently questioned whether this ideological binary determined the form of black higher education in the United States in any widespread way. It is a debate that may not have had much direct expression in most black colleges. Possibly, the tension between "practical" or trades training and "classical" preparation of black youth for more professional occupations held institutional significance only in the few settings where considerable resources might be expended on education and educators had the opportunity to choose their curricular course. Most public black colleges functioned with so little reliable funding that long-term pedagogical strategies of any kind found little application. Relatively stable private HBCUs, run by philanthropies, adhered to set philosophies determined by their patrons. In short, curricula were contingent on the changing demands of sponsors or constrained by a lack of sponsorship altogether. Certainly throughout World War II, instruction at Eastern Shore was little more than a placeholder for black higher education in Maryland. It really functioned as a secondary school; the only one outside of Baltimore to offer high school equivalency opportunities in reading, writing, and math for black citizens but nonetheless, offering an educational opportunity that did not equate with those offered to white college-age students in the state.[9]

Despite the fact that federal monies flowed to land-grant systems on the basis of white and black state populations, all seventeen segregated land-grant systems of the early twentieth century provided deeply discrepant services to their states' black populations. However creative and energetic faculty and administrators of the black land-grant institutions may have been, there was little hope of concerted development for most of these schools. Trends that built up science and engineering at white land-grant schools had little impact on black campuses. When the federal government perceived a need for greater support for farm and home extension services offered by the nation's white land-grant universities, the resulting Smith-Lever Act of 1914 provided no funds expressly for the black land-grant schools. It left the choice of how much to spend on Negro land-grant colleges up to their white counterparts in each state, resulting in a predictably low level of appropriations for the black schools. The first third of the century saw a marked "scientization" of agricultural research and teaching in many white universities, but no state then operating a black land-grant college set up an Agricultural Experiment Station at its black campus despite a trend to create such stations at the white campuses and the rapid growth of revenue from those facilities. At Eastern Shore even extension work was minimized, denying black citizens an avenue toward advanced agricultural science work or exposure to rural engineering, economics, or sociology.[10]

By the end of World War I, the issue of how extensively and with what resources the nation should educate its black citizens rested on ideas of how best to fit minorities into a modernizing industrial economy. Wartime economic expansion and new roles for industrial research and development vitalized technical programs at U.S. universities. At the same time, northern and urban settings were the initial locus for increased agitation for improved black participation in higher education, and the federal government in certain regards supported that shift. As John Thelin documents, black college and university attendance overall was growing: HBCU enrollment grew from around 2,000 in 1918 to 14,000 by 1930. Eastern Shore also experienced some changes. Instruction at the level of grades eight through ten was eliminated in 1927, and two-year college courses were inaugurated; the campus's first four-year degrees, in agriculture and mechanic arts, were offered in 1935. However, in 1928, the state had a black population of around 260,000, of which only 433 were enrolled in college. Federal studies have revealed that this miniscule level of black college participation was fairly typical of the segregationist states, many of which had minimal accommodation for primary and secondary education for black youth that might prepare them for college admission.[11]

During the 1930s, Maryland provided about $26,000 per year to the Eastern Shore campus. Teacher training, mechanical arts, home economics, and agricultural disciplines gradually expanded their offerings. The school's curricula, however, nonetheless continued to match "the mind of the rural teacher," according to one historian. It remained vocational in its focus, honoring few aspirations beyond the limited range of occupations already open to blacks in the Mid-Atlantic states.[12]

The idea that Eastern Shore might undertake graduate or research work was never mentioned during this period. As was the case with nearly all seventeen segregated land-grant systems operating black campuses, such decisions at Eastern Shore remained in the hands of whites. All resources and decisions about the scale and nature of curricula were controlled by officials, whether state legislators or Boards of Trustees, who worked with administrators to determine the sort of instruction offered to African Americans. In mid-1930s America, a white person aged 18 to 20 was four times more likely than a black person of that age group to attend college, and the desirability of including black youth in the most skilled labor pool was by no means clear.[13] Maryland, with considerable industrial growth occurring in Baltimore, faced a more urgent set of questions about such workforce issues than did some of the industrially less-developed southern states. Would Maryland's black citizens continue to fill a stratum of low- and semiskilled production jobs, with their greatest opportunities being in teaching at the primary and secondary level, or would they begin to find a role across a much wider segment of the economy?

For the previous half century, many educators and policy makers concerned with black education had maintained that accommodationism was the optimal position for African Americans: the black citizen must understand her or his place in the emerging industrial order, and that place was near the bottom of the economic hierarchy. Few would deny that the teaching, medical, dental, and law professions would require a regular supply of black practitioners: in any segregated and many integrated communities there was little crossover between black and white consumers in these areas. Few social observers could imagine a world without a color line in these professions. But in industrial and military or governmental spheres, as higher education began to promise a more skilled and specialized national workforce, questions about where nonwhites might fit were being newly articulated. It seemed possible to some observers that jobs in the productive sector, at many levels of skill and training, would conceivably welcome people of color who were thought to show promise.[14]

Even as legal changes began to challenge the historic exclusion of blacks from white educational settings in both the north and the south, black schools faced new versions of old questions about where to seek resources, how to target their curricula, and, in newly articulated terms, what role to give racial identity. The visions of black educators ranged from expectations of nearly equitable participation by minority citizens in all fields of U.S. higher education along existing lines of eligibility that determined white opportunities, to compensatory agendas that maximized inclusion for students of all levels of achievement and preparation. In this latter view, the goal was to bring as many young people as possible into the sphere of higher education. Based on the maintenance of open, or nearly open, admissions, schools in this schema were intended to provide whatever level of education might lead to credentials for the greatest number of young people. This vision invoked a debate about appropriate standards for black colleges that reflected concerns about resources and reputations (to which I will return in later chapters). Clearly, many felt that there was a trade-off between ideologies of inclusion and institutional standing. What is striking, however, is that the choice between selective and open admissions never seems to have presented itself at College Park, UMD's white branch, as any sort of practical dilemma. Selectivity in admissions may have been debated by those concerned with the white public university's status as the school attempted to move from an agriculturally focused to a more technically minded institution through the twentieth century, but the university seems never to have considered restricting services because of limited resources. From its earliest years, the school at College Park could count on the advocacy of Maryland's elite white citizenry and, after Byrd's appointment as president, on legislative support as well to grow as needed and thereby maximize services to white students.

College Park and the Centrality of White Opportunity

College Park opened in 1859 as the Maryland Agricultural College, and became a land-grant school under the Morrill Act in 1864 when the state legislature approved that designation. Some of the nation's first professional schools had started in Maryland in the early part of the nineteenth century, arising from concentrations of wealth in and around Baltimore. But by mid-century, the state's rural "gentleman planters" were worried about their diminishing influence and they sought an outlet for their own cultural aspirations in the new institution. University of Maryland historian George Callcott detects a double function for this early agenda at the state university. Representing the interests of the established elite in the state, the Maryland Agricultural College would provide a site for developing new, science-based knowledge about farming. At the same time, it would be a place that would bring the "common" small farmer and rural masses some training, character, and manners.[15]

The Maryland Agricultural College nearly dissolved at the end of the Civil War, during which its sponsors had largely sided with the south (if not actually advocating for secession by Maryland). However, through the 1860s and 1870s, adjusting to a greatly altered national economic climate, the school began to participate in transitions wrought as northern corporate ideologies overtook agriculturalist traditions in the south. With an increasing focus on industrialization, the state as a whole saw the old aristocratic power structures of both Baltimore and its rural regions shift to a more commercially minded leadership, with the corporate sector gaining political influence. This transition brought, in turn, a greater emphasis on skills and knowledge that would be of value to industry as the college's Board of Regents added members from the corporate sector. In a pattern seen in universities throughout the country after 1900, lawyers, executives, and bankers began to displace the clergy and landed gentry who had once dominated the school's directorate.[16]

This shift to the broader interests of commerce, however, should not be mistaken for an egalitarian trend or for one in which old racial divisions faced considerable opposition. University of Maryland President Raymond A. Pearson, in office from 1926 to 1935, worked with the system's Board of Regents to support the common interests of Maryland's governors, legislators, and influential white citizens. Pearson placed particular importance on building up the school's physical plants at College Park and in Baltimore, but showed far less interest in any kind of systematic academic modernization of the UMD system. If academic programming at College Park held little importance for Pearson, its status at Eastern Shore was negligible. That campus's very low enrollments, and low pedagogical standards even after it began offering four-year degrees in the late 1920s, can be blamed in part on the Depression, which eroded higher education budgets and facilities across the country. But for an understanding of the role of racialist ideologies in shaping higher education, it is necessary to consider some direct measures of the state's minimal commitment to its black land-grant school. First, in 1930, the salary of the president of the Eastern Shore branch, at $2,400 per year, was lower than that at every other southern black land-grant college except that of Louisiana, with which it was tied. Only four schools among the seventeen states with dual systems paid the presidents of their black land-grant colleges less than $3,000 per year; others paid nearly double Maryland's salary. Eastern Shore had no registrar, dean of men, or dean of women; only West Virginia, which paid its president a salary twice that of Maryland, had so small an upper administration at its black land-grant institution.[17]

Perhaps even more tellingly, in 1932, the UMD Board of Regents established a scholarship fund of $600 to support out-of-state study for black college students unable to find suitable degree programs in state. As the Depression worsened and pressure for government intervention in the nation's social welfare increased, progressive interests identified huge discrepancies in black health, housing, and education. The failure of separate but equal provisions to improve black education was now more widely acknowledged; in Baltimore, the NAACP began its first challenge to the definitions of *equal* that had allowed segregation to stand in the courts' eyes. Scholarships for black students were a common means by which segregated institutions hoped to deflect such challenges. The $600 scholarship fund, intended to cover the expenses of multiple students at the undergraduate and graduate level, was pathetically small, but represented a decision to fund scholarships *instead* of improvements at Eastern Shore. Those educators, politicians, and citizens genuinely interested in supporting black higher education in the state tended to reside in Baltimore; when faced with creating a strategy for change, many understandably promoted the expansion of Morgan College over the less developed Eastern Shore campus. To others, the state's two black teachers' colleges (Bowie and Coppin State) seemed more worthy of funds than the markedly underserved agricultural and mechanical school. In fact, few real dollars flowed to any of Maryland's black institutions, but activists in Baltimore were beginning to find new sources of support for race reform, especially as the NAACP gained a following there, and an increasingly charged political atmosphere in the state would soon lead UMD's fretful regents to replace Pearson with a far more politically astute operative: H. C. "Curly" Byrd.[18]

Byrd's World

Harry Clifton Byrd took over leadership of the University of Maryland system in 1935 when Pearson was finally deemed untenable. Pearson's neglect of conventional academic standards and some questionable connections to political interests had gradually undermined his reputation among students and faculty, and, eventually, with alumni and the UMD Board of Regents.

Byrd, born in 1889 to a family of oyster fishermen on the Eastern Shore, had come to College Park as a student at the age of nineteen, and began a decades-long involvement with the UMD system. He graduated with a degree in civil engineering from the University of Maryland, and did postgraduate work in law and journalism at Georgetown and George Washington Universities and Western Maryland College. In 1912 Byrd accepted a position as an instructor of English at College Park, alongside duties as a football coach. As a student his football heroics were acclaimed; he served as an athletic coach until 1932, continuing to make a priority of the university's athletic programs during his years as an administrator. Byrd served as an assistant to Pearson until, upon Pearson's resignation, the Board of Regents asked Byrd to step in as acting president, a position he held for eight months. In February 1936, Curly Byrd was formally appointed president of the University of Maryland system, with responsibility for the College Park and Eastern Shore campuses, as well as for the professional schools in Baltimore.

George Callcott has noted Byrd's powerful but contradictory personality as expressed in a set of exceedingly strong loyalties and prejudices that determined the operation of the university system until Byrd's retirement in 1954. Although Byrd left the post that year to run for governor (a race he lost), it seems likely that by that date he was no longer operating entirely in sync with ideas about public higher education in Maryland. Regardless, for nearly twenty years, he brought an intensely focused if stubborn management style to the job.[19] Famously compassionate in his personal relations, he was closely involved with many individual students, staff, and faculty. His tendency to micromanage is evident in his voluminous correspondence, which includes minutely detailed instructions to groundskeepers fighting pond scum around a sewer outlet at the Eastern Shore campus and lengthy refutations of critical editorials in regional newspapers. Concerned alumni and Maryland citizens, who turned to Byrd as a source of reassurance and action on many aspects of civic life, commonly received personal responses to their letters.

Callcott depicts Byrd, who was often described as handsome and flirtatious, as developing a "cult of personality" while at the same time displaying an "egotism" that irritated many. There can be little question that the cooperative relationship Byrd maintained with powerful business leaders and lawmakers at the state and federal level promoted Maryland's entrenched interests and excluded those who had little voice in public matters. His standing on race remains one that is still debated on the campus, where the football stadium today bears his name.[20] He worked at every juncture to preserve the dual educational system that kept blacks away from College Park, and he tailored his rhetoric to a climate in which arguments against segregation came in many varieties. If overt claims of black inferiority are absent in Byrd's writing, his policies nonetheless preserved a thoroughly biased system. Many of the individuals to whom he offered help and reassurance were white citizens worried about the effects of integration on their jobs or communities. He exerted his influence on behalf of Maryland's white interests to maintain a segregated university system, and economy, in that state.

As Byrd saw his job, he carried the UMD system into line with a national wave of economic expansion and modernization, while preserving an exclusionary social order. Between 1930 and 1945, enrollment in U.S. colleges and universities grew tremendously. As more and more people attended high school, eligibility for higher education increased. Between the two world wars, enrollment in universities across the nation grew from about 250,000 to 1.3 million. As Thelin points out, as more people attended college, observers began explicitly to associate higher education with a kind of collective uplift. Popular media celebrated postsecondary education as a tool of societal and individual improvement: "These boys and girls...will in twenty years occupy the seats of authority." State universities lacked the venerable reputations of Ivy League schools, but as Thelin indicates, institutional pride and alumni loyalty helped build the multicampus state universities to nearly competitive status with many of the older, private East Coast universities.[21] For Byrd, a new legitimacy and civic importance for the University of Maryland must have seemed within reach.

That vision was based on an increasingly instrumental sense of where higher education would fit in the U.S. economy. Like many other university administrators, Byrd knew that close relations with industry and government held a greater likelihood of delivering new resources to his school than did aspirations of a more elite, Ivory Tower identity. Although the latter may have been unattainable for the UMD system in any case, Byrd's loyalties clearly lay with regional manufacturing and corporate interests, not with the elevated cultural spheres such as those that might have supported institutions like Johns Hopkins University.[22] In this climate, engineering held a new importance for the UMD system, and the College of Engineering became one of Byrd's most successful institution-building efforts at College Park.

Engineering had been established as a distinct school at College Park in 1919 during a reorganization that also created schools of Agriculture, Arts and Sciences, Home Economics, Chemistry, and Education, as well as a freestanding Graduate School. Its departments offered degrees in civil, electrical, and mechanical engineering. In the early 1920s, research facilities were established for the Maryland State Roads Commission on the College Park campus, as was typical of land-grant engineering schools eager to serve growing state infrastructures. However, with only about 500 undergraduates and a handful of graduate students at the time that Byrd assumed office, engineering work at the University of Maryland was slow and small scale compared to that being conducted at other land-grant institutions.[23] The state universities of Illinois and Iowa, Pennsylvania State University, and Cornell University by this point had decades of highly reputed engineering research and teaching to their credit, and their faculty and alumni dominated the nation's engineering professional associations. These public schools competed easily with the engineering divisions of private universities, offering vital services and personnel in research and development, materials testing, design, planning, and other arenas of great use to military and productive enterprises. Byrd, once appointed president, saw an opportunity for growth along these lines at College Park, and specifically the chance to compete with the private Johns Hopkins University for the lion's share of state funding for higher education in engineering. Byrd soon hired S. Sidney Steinberg as College Park's dean of engineering. A civil engineering graduate of Cooper Union, Steinberg had experience with state highway departments and in hiring him Byrd passed over at least one other candidate who had a background in agricultural engineering. It seems likely that Byrd intended Steinberg to craft the enhanced connections to Maryland's commercial and governmental spheres that the president so eagerly sought.[24]

The Growth of Engineering at College Park

In hopes of receiving accreditation for the engineering college, Steinberg almost immediately invited the Engineers' Council for Professional Development (ECPD) to examine the engineering curricula at College Park. The ECPD's Committee on Engineering Schools, then headed by Karl Compton out of MIT, found engineering at College Park to lack some basic elements across civil, electrical, and mechanical engineering and in such vital foundational areas as physics. According to Compton's report, laboratory facilities were in some areas "very meagre" [sic] and faculty badly overloaded. Byrd and Steinberg quickly worked together to arrange for new equipment, facilities, and staff to answer these criticisms. Byrd persuaded the Maryland legislature to provide $150,000 in early 1937 for this purpose, and College Park's engineering curricula were accredited by the ECPD before the end of the year.[25]

Interestingly, Compton's report had pointed out that the University of Maryland offered more courses in Municipal Sanitation than was "customary." Steinberg, however, disagreed and made the case that "this field of work is very important to the development of the State of Maryland." With the dean and a growing engineering faculty, Byrd was seeking new connections to industrial and government sectors that might form a sort of patronage network for engineering at College Park. He arranged for the establishment of research facilities on the campus for the U.S. Bureau of Mines in 1938, and then for trade groups that included the National Sand and Gravel Association and the National Ready Mix Concrete Associations.[26] Despite their mundane names, these sorts of organizations actually represented significant connections to industry. For many engineering schools, trade organizations of this kind provided a steady source of paid research work for university engineering laboratories, brought welcome publicity to faculty, and led to employment for many graduates. With a sense of the broad utility that this kind of academic program might hold for the state, College Park also began in this period to conduct short courses for waterworks and sewage plant operators and for engineers of state roads. Byrd was shaping the engineering school at College Park in the mold already familiar to older and more established land-grant programs. Validating his own strategy, he summarized his efforts in this direction as "horse sense" and penned a lengthy account of these new programs to the *Baltimore Sun,* a newspaper already beginning to stake out a position strongly opposed to Byrd's leadership. Finally, in 1937, Byrd and Steinberg oversaw the creation of an Engineering Experiment Station that greatly expanded UMD's potential for involvement in industrial research—a relatively late but necessary development for the land-grant if it hoped to reach national standing in engineering fields.[27]

Byrd was at every turn a keen observer of emerging markets for technical expertise. In 1937, he brought Wilbur J. Huff from Johns Hopkins to head a new Department of Chemical Engineering at College Park. That program initiated the second master's-level program in the College of Engineering (Electrical Engineering having established the first in 1926), and the college's first PhD program in 1939. This growth likely held multiple strategic meanings for Byrd, whose competitive feelings toward engineering at Johns Hopkins colored many of his decisions. Hopkins, which had had a highly reputed engineering program for decades and close relations with regional industries, at this point sought a designation as Maryland's "state school of engineering," which would have deprived College Park of important legislative funding and thus the hope of real prestige in technical fields. Hopkins' civil engineering program was much larger than that at College Park, and its mechanical and electrical engineering programs were well established. It offered 336 semester-hours per year in engineering, to College Park's 198. Undaunted, Byrd publicized his hiring of many faculty for College Park who held PhDs in engineering fields from Hopkins, as well as a number of professors who had previously been employed at the rival institution. He was in a sense dependent on Hopkins as source of expertise, but saw that in luring its personnel away to his own institution he might weaken that university as he benefited his own.[28]

In pursuing accreditation, the UMD College of Engineering added more hours for its engineering degrees and expanded the requirement for senior theses to electrical and mechanical engineering, in addition to civil engineering. By implementing the addition of fundamental mechanics coursework, including statics, dynamics, strength of materials, and hydraulics, Byrd attempted to align the program with what he saw as the "standard in all engineering colleges." Byrd also showed a keen interest in work by the Society for the Promotion of Engineering Education that gave equal billing, if not equal classroom time, to "the sciences that deal with human relations" alongside scientific and engineering subjects. In so doing, he echoed contemporary claims for the suitability of engineering graduates for managerial roles in industry and government. Since shortly after the turn of the century, professional engineering groups in the United States had worked to claim this kind of administrative status, and university engineering schools were sites of significant self-promotion along managerial and executive lines. Byrd seems to have been well aware of this trend, which continued to grow as technical expertise became more and more specialized as the century progressed. He resolutely kept industrial training, geared toward vocational and educational occupations and unlikely to lead to managerial employment for graduates, separate from engineering, understanding that such vocational training could co-exist with a prestigious engineering college if the two were entirely distinct in function and reputation. Following such established practices, he crafted a credible and competitive new engineering school at College Park.[29]

Technical Instruction at Eastern Shore

As Byrd developed engineering programs at College Park, he systematically denied the need for any such agenda at Eastern Shore, a setting he saw as remote from UMD's main branch both geographically and conceptually. At every juncture, arguments about the role of universities in scientific and commercial development, so freely deployed in Byrd's aspirations for the College of Engineering at College Park, were elided or denied in discussions of the Negro branch. Even comprehensive vocational instruction, justified at College Park because it served the labor needs of regional industry, found little purchase at the black campus. Instead, through the late 1940s, Eastern Shore offered its students a narrow range of majors, and those were provided in under-staffed and under-equipped departments. Black Marylanders were seen, at least from the vantage point of College Park administrators, as ineligible for meaningful opportunities in higher education and for any prospect of contributing economically to the polity.

This official dismissal of black economic opportunity and contributions was apparent in the physical plant at Eastern Shore. In the mid-1930s, the black campus occupied some 200 acres, of which 170 were under cultivation. The site held relatively few structures, and many of these were decrepit. An administration building with six classrooms, a frame building for classes in mechanic arts, a library, and a handful of barns and other utility buildings stood beside two rundown dormitories. Faculty and administrators had few resources with which to work, and when the school increased the length of its courses of study from two years to four years in 1935, no plans were made to add buildings or equipment. Of its eleven faculty members teaching in the 1936–1937 school year, only three held master's degrees. At this time, capital investment at College Park amounted to more than $4 million, spending on the professional schools in Baltimore was measured at almost $2.7 million, while Eastern Shore showed buildings and equipment amounting to a value of only $100,000.

In 1937, Maryland's State Commission on Negro Education found these facts to show "with striking force the failure of the state to provide a state institution for higher education of Negroes in any way comparable to what has been done for white students." With only forty-six students attending the neglected campus, the commissioners suggested closing the school and allowing Morgan College, valued at nearly a million dollars at this time, to take over its functions at Morgan's Baltimore campus. They understood that Eastern Shore poorly served those students who did enroll there, functioning as a token attempt at black land-grant education that could forestall more ambitious development by fulfilling the letter of separate but equal laws. If the Eastern Shore facility could not be abandoned altogether, the commission recommended converting the site back into a secondary school. Instead, in part due to Byrd's insistence, the state chose to continue the existence of the school at Princess Anne. The state formally purchased the facility from Morgan College, changed its name to Maryland State College, and began a decade of adversarial relations with the many supporters of Morgan who had hoped to see the state's resources instead consolidated on a single, well planned urban campus in Baltimore.[30]

This strategy of building up the old facility at Princess Anne could be interpreted as reflecting simple careerist self-interest on Byrd's part: annexing the college at Eastern Shore would add to his administrative dominion. What is more, for those in the Maryland legislature who hoped to see the Eastern Shore campus built up, its closure could have been construed as a personal failure on Byrd's part. Surely Byrd would not have wanted to disappoint his allies. However, Byrd's commitment to maintaining a segregated university system suggests that he saw the Eastern Shore campus as a means of closely controlling black educational opportunity in Maryland. Morgan College had historically operated under the influence of powerful political and civic interests in Baltimore. By maintaining the Eastern Shore facility, and extracting its operation from Morgan's control, Byrd could direct a good portion of state spending on black higher education, as he had already managed to do with Maryland's white public higher education at College Park. In this way, however small that funding for the black land-grant might be, Byrd could maintain his political relationships while enacting social controls in an increasingly unsure atmosphere regarding race. Pressures were mounting on the UMD system to end segregation. A sign of conservative concern about this pressure was the legislature's decision in the mid-1930s to increase the scholarship fund for out-of-state study by black students from $600 to $30,000 per year. This allowed more black students to undertake "equal" undergraduate or graduate work without seeking admittance to the state's white institutions. In the 1938–1939 school year, Maryland supported 158 African American undergraduate and graduate students with scholarships of this kind. Some black citizens objected to the scholarships ("the mere closing of its doors to any group of taxpayers stamps the Maryland university as a cheat and a robber"), but Byrd had no intention of capitulating to their concerns. Byrd's individual policy decisions about black education in Maryland are not surprising in light of his claims in the late 1930s that "perhaps I shall have to go to jail, but I think we have got to keep the Negroes out" of College Park, and that a failure to maintain that color line "will come pretty close to ruining us."[31]

Obviously, no such defensive claims would have been necessary on Byrd's part if challenges to racial inequities were not being detected by conservative Marylanders. Such challenges were not always or even commonly directed against segregation. More usually, reformers sought greater resources for Negro institutions. In the 1930s, Maryland expended less on black higher education than almost any other border or Southern state (around 36 cents per Negro inhabitant in the 1935–1936 fiscal year; Delaware, allotting $16 per Negro inhabitant, spent the most among these states).[32] Byrd was instrumental in keeping such disparities in place. As the president of the state's land-grant university system, he helped define a categorically different kind of economic role for Maryland's black citizens—one predicated on African Americans' rural residency—and, commensurately, a need for agricultural and perhaps mechanical training, but not for more advanced technical skills such as those represented by the sciences and engineering. While blacks living in Baltimore might work in industrial and other commercial jobs, the land-grant system did not need to serve that demographic group or provide any possibility of unlimited economic mobility to the rural black citizens it did consider to be part of its constituency.

Faculty and administrators at College Park envisioned mechanic arts evolving separately from engineering in order to maintain the prestige of the latter. Black land-grants commonly undertook no engineering whatsoever, indicating their complete removal from the more prestigious fields of technical teaching, research, and commerce. As David Wharton's account of black entry into U.S. engineering programs indicates, by the late 1930s, Howard University, North Carolina Agricultural and Technical College, and Hampton Institute had established engineering programs for African Americans. Howard had built its engineering programs with faculty drawn from the small number of black graduates of integrated engineering schools in the north, including MIT, but its tuition was unaffordable to most southern blacks. North Carolina A & T was the only land-grant among the three schools, and a glaring exception among the black land-grants in its commitment to engineering.[33] Instead, as Chapman's 1940 study reports, technical instruction at the black schools centered on mechanical drawing, automobile or tractor mechanics, building and power plant construction, printing, and other industrial skills, as well as training for secondary teaching in any of these fields. Eastern Shore offered no programs in mechanical, civil, electrical, or general engineering; subjects that were covered by some sixty courses per year at College Park in the late 1930s. "Drafting" constituted the only subject matter listed under "Engineering" rubrics at College Park also to be offered at Eastern Shore.[34]

Byrd's association of African Americans with rural agricultural sectors of the state's economy represents his linkage of social and knowledge organizations. In crafting this connection, Byrd's ideology meshed with that of some of the most prominent experts on black education of the day. Among these experts were many proponents of a practical emphasis in higher education, but what that practicality might mean differed depending on the race of the population to be educated. Rhetoric about curricular content could change its meaning according to context. For example, engineering majors at College Park were encouraged to attend lectures by "prominent practicing engineers" so that they might grasp the day-to-day nature of different technical disciplines. Like students at virtually every U.S. engineering school, they were also given laboratory and field experiences meant to cultivate hands-on appreciation of engineering processes and materials. This was not so much a matter of improving manual dexterity as immersing engineering majors in a workplace culture that associated subjectivity and discretion with managerial potential. At majority engineering schools, working with one's hands carried an ascription of physical strength and manhood more generally. Engineering pedagogy from 1900 onward stressed the fortitude required for accurate, efficient technical work and university curricula cultivated that pairing of mental and physical discipline. By contrast, in the case of minority students, a practical pedagogical focus did not mean preparation for the production conditions of industry, let alone for the fulfillment of managerial aspirations. Rather, it meant centering course offerings on lower-skilled agricultural or industrial training that critics have identified with accommodationist racial theories. Here, getting one's hands dirty held few romantic implications of heroic physicality. If engineering students at College Park gained a vital sense of their own manly capacities by shoveling gravel, at Eastern Shore shearing sheep, milking goats, and making mattresses instead projected the upward limits of black vocational potential.[35]

The "Agricultural Negro"

Byrd's advocacy of a rural, agricultural lifestyle for Maryland's black citizens probably seemed reasonable to many anxious whites. With Baltimore at the center of race activism, to bolster Morgan College as the main site of black public education might have stoked the fires of civil unrest. What is more, Byrd could build on a set of arguments about the salutary nature of rural life that had a long history among southern educators. By associating those benefits with Maryland's black citizens, he might not dismantle the thriving (and politically resistant) black communities of Baltimore, but he could discourage the movement of rural blacks into the urbanized commercial sectors.[36] As the concerted development of a technologically sophisticated workforce in the south gained importance after the 1920s, this kind of logic confined those new occupational opportunities to whites. It effectively encapsulated workforce needs to prevent them from bringing about an erosion of racial structures. But it is only by understanding positive attitudes regarding education and technology in southern economic development that one can appreciate the extent to which the pastoral black identity at work in Maryland reinforced racist projects in that region.

For many years, historians understood southern culture to be averse to modernization, seeing in the rural bases of its economy an expression of cultural disinclination toward industrial development and its associated educational projects. In the nineteenth century many elites from the planter class did look with limited enthusiasm on those priorities and, putting aside the somewhat artificial periodization of "Old" and "New" South (Emancipation obviously did not alter every white southerner's worldview), one can see a region slower to adopt urban, industrial development than many northern settings. Yet long before the "Americanization" of the South, as the creation of Sunbelt industries in the 1970s and 1980s has been labeled, influential southerners understood the importance of science and technology to regional affluence. World War I did much to dissipate what remained of academic and corporate disinterest in research and development; by the 1930s, Byrd was enthusiastically expanding engineering at College Park, as we have seen.[37]

The administrators and trustees of the UMD system imagined no such upgrade occurring at Eastern Shore. By the 1920s, the model of white education that stressed vocational training, farm demonstration, and homemaking as integral to traditional social and moral values had given way in the south to the goal of more varied education for majority citizens. But for segregationist educators like Byrd, that rural model held enduring utility for black education, helping delineate blacks as a group meant to remain "pre-modern." In an era of rapidly growing cities characterized by an increasing ethnic diversity, those hoping to consign black citizens to rural regions could draw on a strong mythic tradition that praised rural life as a source of social stability, and not incidentally, of "moral and racial purity." White culture in the south may have been rapidly facing the expiration of this avowed agrarian identity, but its language remained powerful and bolstered the sort of gradualist views that saw limited technical or corporate involvement for blacks.[38]

Byrd received advice from many businessmen, lawyers, and other civic figures in Maryland along these lines. In a letter from one correspondent, which Byrd kept in his "Negro Education" file, a set of fairly typical ideas for the "betterment of the Negroes of Maryland" appears. Baltimore attorney William L. Fitzgerald, at one point a member of the Baltimore City Council, took an active part in official and ad hoc citizens' groups addressing race relations in Maryland. In a treatise that Fitzgerald prepared for Byrd, the lawyer called on the public university system to outline for Maryland's black citizens the "advantages of rural life." Fitzgerald explained that blacks of the rural regions might find that supervised recreation, improved personal hygiene, and improved saving habits ("dependable industry, progressive living, and thrift") would reinforce one another. The presumption that this population would benefit from instruction regarding health and finance clearly implies condescension. Although black morbidity and mortality rates were higher than white rates during this period, compare Fitzgerald's behavioral prescriptions for improved health among minority citizens to alternatives such as state provisions for improved medical care, or farm subsidies and job training as means of raising rural incomes. The lawyer's advice was undoubtedly in sync with Byrd's equation of segregation with an orderly populace. Fitzgerald concludes his letter with the point that these innovations, including urging Negroes to "stay on the farm and make it pay," would make Maryland a "better and safer place to live."[39]

The steady migration from south to north and from rural to urban communities, and the increased availability of industrial jobs, was creating a pattern of black movement from agricultural to industrial employment in many states that made a strictly agricultural economic role for blacks seem outdated. But the UMD system forestalled any "interaction of identities" between its white, research-oriented campus and its black vocationally focused school. Fitzgerald's blunt endorsement of a racialized economy was reinforced by its underlying premise that the black farm economy functioned separately from the industrial sphere; this was patently not the case in emerging white economies of the 1920s and 1930s, in which agriculture itself became much more of a consolidated and mechanized enterprise, with stronger ties between individual farmers and commercial supply and distributions systems.[40]

New Deal policies intended to help a deeply troubled national agricultural sector revived old populist ideologies about farming as a source of democratic energy and idealism in the nation. Thelin notes that the University of California's School of Agriculture in Berkeley adorned its 1917 building with the motto, "To rescue for human society the native values of rural life," and such exaltations were common through the Depression.[41] But the myth of the yeoman farmer was a white ideal, never intended to improve the conditions under which black farmers lived in the United States. While it unquestionably brought economic security and improved production to many locales around the country, the entire land-grant movement was in some ways equal parts commercial scheme and romantic vision. Populist sentimentality endured into the age of agribusiness—the consolidation and industrialization of farming—a process to which many land-grant researchers contributed their efforts.

What is more, in southern and northern settings, agriculture and scientific modernization were far from mutually exclusive: the application of science to agriculture was a huge part of white land-grant university activity before and after 1900. But the benefits of science-based agricultural research flowed toward the relatively few farmers who could afford to implement innovative practices through the capitalization of mechanization or new stocks. Many farmers in the United States operated on a small scale with little margin for expansion or experimentation. Many southern whites in agricultural occupations, in particular, fared poorly in this era. However, although a significant proportion of white farmers had little hope of upward mobility, sharing that constraint with black farmers, it was only whites who were given the chance of attaining the full range of corporate and industrial opportunities associated with economic modernization, particularly in higher-tech and white-collar sectors. From the 1880s onward, skilled industrial employment in southern states went increasingly to whites; even the textile industry, by far the largest in the region, employed relatively few blacks and only at the lowest levels of technical training.[42]

Education reinforced this difference. Since the 1890s, southern school reformers had advocated that only a greatly expanded system of public education for the region's white citizens would assure the economic development of the south; black education seemed to have little place in this logic well into the twentieth century. Even the most culturally "backward" white communities, as reformers labeled parts of the Appalachian South, were eligible for this kind of uplift, while African American education, by contrast, received attention only from northern philanthropic organizations that could never reach the scale of state-supported efforts for white communities.[43] What is more, agriculture at the black land-grants rarely functioned at the level found in the white state schools. By 1890, over 25 white land-grants maintained Agricultural Experiment Stations or equivalent facilities. For the first half of the twentieth century, only Tuskegee among black land-grant schools had an Agricultural Experiment Station; it was here that George Washington Carver did his research, entering as founding director in 1897. Prairie View A&M University in Texas opened its station in 1947, but the next Agricultural Experiment Station at an HBCU did not open until 1971, at Mississippi's Alcorn State University. Through all of these decades, black farmers in the south received a fraction of the benefits in expertise and assistance to which white farmers had access. Remarkably, it was not until 2007 that the federal government required states to match federal support for research and extension at black land-grant institutions at a rate of 100 percent.[44]

In spending far less on Eastern Shore's agricultural extension programs than was spent at College Park throughout the first decades of the twentieth century, the UMD system made it clear that black farmers were of a different caliber than its white ones. As Callcott recounts, the agrarian reform impulses that had motivated the University of Maryland's founders were predicated on the idea that education would uplift the "poor farmer to skilled agriculturalist" (and raise the "blacksmith and stonemason to engineers"). Yet in 1937, when blacks constituted nearly 18 percent of the state's rural population, only 2.3 percent of the state's expenditures on rural extension work was deployed for work with black communities.[45] The Smith-Hughes Vocational Act of 1917 channeled federal money to vocational training but, like the Smith-Lever Act, did not specify that any of that money need go to Negro land-grant schools. Each state operated according to its own discretion, and Maryland, alone among the southern states, designated none of its Smith-Hughes funds for black education between 1928 and 1938. As Chapman calculated for his 1940 study, equitably distributed, aid for land-grants from the Smith-Hughes Act in the 1937–1938 school year should have provided 17.8 percent of that funding to Eastern Shore, amounting to around $17,000. Instead, the state channeled only $1,900 to that campus. As Chapman dryly summarized, "It seems to be the practice in Maryland to count the Negroes in order to receive Federal money, but to forget them when the money is disbursed."[46]

Byrd, closely involved with the flow of funds and influence among Washington, Annapolis, and College Park, surely had knowledge of these disparate expenditures and it is nearly certain that he approved them. Even those fields in which the Eastern Shore school was relatively well established, as in the training of teachers for vocational and agricultural subjects, had little stability. Public education for Maryland's black citizens was underfunded at all levels; without considerable numbers of high school programs in those subjects, trained teachers were unlikely to find employment. Maryland reported an oversupply of teachers of vocational agriculture in a 1938 Harvard study that implied not the existence of too many teachers, but of too few students.[47]

Prior to World War II, black communities outside of Baltimore had relatively little legislative representation compared to the white majority, and it is clear that the state legislature gave low priority to the Eastern Shore campus. This stands in contrast to the legislature's support of Byrd's vision of a much expanded role for technical education in College Park, serving all levels of industrial research and personnel requirements. Chapman summarized in 1940 the many conditions that stood in the way of engineering programs at Negro land-grants. He described the situation as an "intricate educational problem," and the heterogeneity of obstacles to engineering is telling. For one thing, manual training, trades, and machines-shop practice could not easily give way to programs in electrical, mechanical, civil, and chemical engineering if no budget for equipment and new physical plants existed. But Chapman also points to the difficulties upstream and downstream of that fiscal limitation. The problems of securing the black engineering faculty believed to be required for the black land-grant schools and of finding employment for black engineering graduates both discouraged efforts to create engineering programs at HBCUs.[48] For the first decade of his presidency, Byrd showed no intention of bringing an infusion of funds to Eastern Shore to disrupt this pattern, even as reformers found the first footholds for the dismantling of *de jure* segregation.

An "Unpleasant Notoriety"

The first of the major legal developments seriously to threaten segregation at the University of Maryland emerged in 1933. That year, Donald Murray, a graduate of Amherst College, applied to the University of Maryland to study law. President Pearson turned him away, suggesting that he apply instead to Howard University. The idea that Howard would enroll black students from Maryland later served Byrd as well, as a sort of regionalized answer for southern states facing the need to establish separate but equal graduate facilities. At the time of Murray's application, Pearson pointed out to Murray that attending Howard would be "cheaper" for Murray than would be enrolling at College Park. The UMD Board of Regents backed Pearson, who throughout his presidency had supported retrograde racial policies and the exclusion of any groups, including the YMCA, remotely associated with progressive causes. At the end of 1935, Byrd, newly serving as acting president, carried the battle forward. It was this case that led him to say he would rather go to prison than integrate UMD's white facilities. Nonetheless, in 1936 the U.S. Supreme Court ordered the University of Maryland to admit Murray to its school of law in Baltimore, citing the absence of any other equal accommodation for black students.[49]

While Murray's attendance in the white confines of the UMD law school was probably a shocking adjustment for the school's leadership, it seems unlikely that they saw this as an isolated incident. As assistant special counsel to the NAACP, Thurgood Marshall wrote to Byrd in early 1937 to caution him about extending scholarships for black Marylanders in lieu of admission to College Park. Marshall felt that Byrd and the UMD Board of Regents were trying to circumvent the Murray decision. He called the scholarship initiative a "high-sounding provision to attempt to exclude Negroes from the University of Maryland."[50] Even groups less urgently pressing for integration but seeking an enlarged scholarship fund, such as one group calling itself "Maryland's Colored Democracy," said that if further support for black higher education was not forthcoming, "there are likely to be unpleasant experiences which will parade the free state of Maryland before the nation as a State which neglects in an educational way approximately one-sixth of its citizens. We have such a pride in our great State that we do not desire to have such an unpleasant notoriety."[51] While lacking the authority of the NAACP, it seems likely that such groups could have embarrassed Byrd, especially among legislators and UMD alumni with connections to a growing black political voice in Baltimore. Byrd's vocal distaste for media and other public outlets of criticism imply that, if nothing else, he was personally disturbed by the group's censure.

A decision later in the decade brought even more profound anxiety to Maryland's segregationists. In December 1938, the U.S. Supreme Court ruled that the University of Missouri must either admit Lloyd Gaines, an African American applicant, to its law school or that a separate school of law be established at historically black Lincoln University. This decision precipitated a crisis among the heads of segregated southern university systems. It was now necessary to provide separate but equal graduate education across the full range of academic disciplines, an extraordinarily expensive prospect.[52] Within weeks of the *Gaines* ruling Byrd was writing to other college presidents faced with reconciling segregationist state constitutions and integrationist federal rulings. He asked the president of Alabama Polytechnic Institute what was being done at that school to meet the "exigencies created by the Missouri decision," reporting that, "frankly, we have an exceedingly difficult situation in Maryland." Byrd felt that the UMD system would "not have a great deal of difficulty" in providing separate but equal facilities for blacks except in "Engineering, Medicine, Pharmacy, Law and Dentistry," a rather large set of exceptions. The solution Byrd envisioned was collaborative: he hoped to enlist the other southern presidents in making Howard University the "State university for professional education for practically all of the Southern states." This strategy openly pitched the ideal of segregated higher education as a common interest among white southerners, and the regional solution addressed both fiscal impracticalities and a perceived threat of eroding values.[53]

Several of Byrd's correspondents responded with enthusiasm, endorsing his idea of centralized black professional schools for the south. Many respondents fretted about *Gaines*; their reactions to the ruling reveal underlying presumptions about racial inferiority that help to explain how segregation remained so entrenched. A few, such as J. W. Calhoun, the president of the University of Texas, suggested that Byrd need not worry that blacks would attend the white campuses if no separate but equal facilities were provided. Calhoun felt sure that "this alternative will never be used."[54] Harmon Caldwell, president of the University of Georgia, specified that not only would the creation of separate black professional schools, including one for engineering, be a "tremendous financial burden" on the state of Georgia, but that no actual demand existed among black Georgians for such facilities: "The number of Negroes in Georgia who would wish to go to the professional schools would be very limited."[55]

Caldwell was grateful, he added, to have Byrd's leadership in this situation. But despite such endorsements, Byrd could not deny a mounting threat to segregation. As College Park eyed new participation in a world of corporate opportunities for research and graduate employment, it could not

Identity and Uplift — 45

avoid increasing contact with more diverse political interests. Fueled by national civil rights trends, the pressure for separate but equal educational provisions for black Marylanders threw the established "low-tech" identity of the Eastern Shore into crisis. The 1940s brought the especially dramatic scientific and industrial expansions of World War II, followed by a post-war world that drove Byrd to ever more extreme positions against encroaching race reforms. However, as I will make clear in Chapter 3, he had many sources of support, both direct and indirect, for his discriminatory racial vision.

Conclusions

Byrd believed that the integration of Maryland's public institutions would create a chaotic social situation with unforeseen consequences for both races. For Byrd, the role that higher education might play in maintaining social order was clear and extended across all disciplines. Of course, white students and black students did not have access to the same disciplines in the institutions dedicated to their separate education. Blacks were systematically denied opportunities in science and engineering, except in the rare cases where they were able to gain admission to out-of-state schools and then obtain funding, either through Maryland's scholarship system or that of the college to which they were admitted. We can accept this denial of opportunity as a project of Byrd and his racist entourage, but we also need to understand that it fit with a broad spectrum of U.S. culture. Across a wide range of racial ideologies, the absence of black citizens from scientific and technological arenas seemed to make sense.

First, on the broadest level, trends in national governance supported segregation, as Chapman summarized in 1940: "The Federal Government acknowledges its obligation of interest in this sad educational plight of Negroes but at the same time feels that it must not make any suggestion to a given state to remedy the educational ills of its suffering Negro group."[56] Ideas about modern knowledge and the differential abilities of white and black persons to understand its value were also at work here. Even groups that found higher education for blacks in Maryland to be profoundly inequitable put forth the idea that blacks were "different" from whites, on multiple levels. For example, as the Maryland Commission on the Higher Education of Negroes reported to the governor in 1937, blacks would tend toward rural lifestyles, given the choice.[57] Chapman himself, amid his indictments of severe discrimination in publicly funded higher education for blacks, launched a scathing attack on an entire category of persons which he labeled "the thriftless and vicious class of Negroes." Based on his own fieldwork among poorer rural families in Maryland in the late 1930s, Chapman seemed to believe that the faults of these black citizens were manifold: he recorded his impressions of poorly maintained homes, ill-mannered children, inappropriate behaviors in public settings, and a tendency toward dissolute lifestyles. Girls of this class, Chapman found, left home at an early age, "lured away by the love of flashy dress, the dance, and travel," or worse still, fell "victim to the glare of the red lights." Boys tired of parental restraints and longed for independence and "money to spend on themselves." His observations led him to judgments about his subjects: in each case, blacks were portrayed as having had choices, and as having made poor ones. This is clear from Chapman's scalding criticism of blacks who did not understand the benefits of modern science and insisted on using "home remedies." Such persons generally failed to understand what a well-constructed, well-ventilated house was, and foolishly feared treatment by doctors and in hospitals. In Chapman's understanding of impoverished Southern blacks, one's intellectual shortcomings were indissoluble from one's social and economic standing.[58]

The idea that modern knowledge and "proper" social behavior are linked helped reinforce Chapman's ideas about how best to correct black exclusion from prestigious occupations. Chapman never underplays the "social restrictions" faced by minority citizens, and refers to discrimination in housing, education, and daily matters of commerce or transportation as no less than "the terrorism in vogue" among white southerners. Unlike Fitzgerald, the white lawyer who advised Byrd on the advantages of "country life" for Maryland's blacks, Chapman acknowledges that the exigencies of the U.S. wage labor system and agricultural tenancy led many black farm families to a lack of stable social or home life. But Chapman, like the white Marylanders to whom Byrd listened and to whom he often appealed, believes that ignorance and willful resistance to change cause some portion of the problems faced by blacks in the United States. Chapman offers, as a contrast to underachieving blacks, this image: "While it is impossible for them to control the range of their income and their occupation, there are certain groups of [black] individuals within almost every income level who have some margin of resources with which to attack these problems."[59] His complex recipe for black economic achievement includes opportunity, values, and self-discipline and thus projects a distribution of blame and responsibility that involves both black and white citizens.

On the most basic level, Chapman shared with Byrd and other southern segregationists a sense that it was through the expansion of formal education that the region's civic and economic interests would be most effectively served. Chapman's faith in the uplifting effects of knowledge of course diverged from segregationists' simply in his imagining that African Americans might develop intellectual sophistication given the proper instruction. But further differences become clear with the creation of separate and ostensibly equal colleges in southern states through the 1940s, as the rest of the nation pressed those states for an end to legal segregation. Based on his constructive suggestions for improving black opportunities in education, Chapman might be presumed to have understood science and engineering as universalizing knowledge systems that would bring common wisdom, and productive outlets, to all who partook of them. The segregationists held no such beliefs. If they had, the creation of separate institutions for black and white higher education would not have made sense; a socially bifurcated body of knowledge cannot be a universal one. These tensions, between modernizing ideologies of expanded technical knowledge and intractable racial bias, are the subject of the next chapter.

CHAPTER THREE

The Disunity of Technical Knowledge

*Constructions of Racial Difference in
Separate but Equal Engineering Education*

FOR MANY CULTURALLY MINDED AMERICANS celebrating the end of World War II, the Allied victory over fascism gave great force to arguments linking science and democracy. U.S. contributions to wartime successes seemed to attest that a nation mobilized for the deployment of science and technology could make major material strides, as measured in new inventions and unprecedented levels of productivity. Through such advancements, any nation might achieve military and economic domination, but interpreted retrospectively through the lens of Allied conquests over dictatorships in Europe and Asia, U.S. commitments to science and technology seemed to assure the triumph of democracy over darker human impulses. This modernist vision was brought to bear on many moral and political projects of the postwar period. Perhaps one of the most famous of these was Vannevar Bush's 1945 government report, *Science—The Endless Frontier,* which was intended to justify the founding of the National Science Foundation. Some science enthusiasts of the day, like Arthur Compton, proposed a positive social side even to the development of atomic weaponry. This dramatic "technical advance" would "force human society into new patterns" that included "greater cooperation . . . the very lifeblood of a society based on science and technology."[1]

As Compton's claim exemplifies, causality was left a bit blurred in many of these optimistic declarations about links between science and democracy.[2] Left unstated in many instances was the precise nature of policy interventions in education and research funding that might bring about this idealized science-based democracy. Nonetheless, the equation of technical and social progress was also adopted by postwar educators and activists anxious to address the most obvious lapse in U.S. democratic values: the appalling robustness of racial segregation in much of the south and racial discrimination throughout the country. Many groups concerned about racial discrimination believed that anti-fascist sentiment could translate into democratic reform on the home front. One spokesman of a wartime labor council hailed integrated training facilities as "the essence of the whole war against Hitler and his theory of race superiority."[3] For some, science and technology seemed to promise especially reliable instruments of social change. The steady industrialization of the southern states over previous decades had set the stage for wider black economic participation, with many African Americans moving from rural into urban occupations, albeit toward jobs at the lower end of the national wage scale. Further, labor crises during the war had "given the Negro greater opportunity" in administrative and scientific fields than ever before, and thus unprecedented chances to display "individual merit and ability." As one prominent educator put it, "The war years of the 40's made ridiculous the old notion that one's race or class is superior to all others and demanded the exhibition of technical skills for the preservation of the nation."[4]

That new demand for a racially diverse technical labor force, however, was rather weaker than it might have been. Despite wartime gains in black industrial employment, supported by the federal Fair Employment Practice Committee, many race activists were far from assured about the inevitability of U.S. societal advances. Some bluntly contrasted Roosevelt's uplifting anti-Hitler rhetoric with continued legal and customary segregation in the U.S. military and virtually every other employment sector. One African American commentator asserted in 1948 that "the peace for which we fought" had simply not been realized, a difficult point to refute when southern colleges that did succumb to integrationist pressures in these years could erect wooden railings to separate black and white students in their classrooms. The GI Bill brought many working-class Americans into university science and engineering programs but did little to dislodge racial barriers. With segregation still in place, and overcrowding and underdeveloped programming endemic at black colleges, black veterans did not even approach the level of opportunities encountered by returning white soldiers.[5] The growing civil rights movement pointed to the incompatibility of democratic political ideologies and persistent Jim Crow practices. Such voices reveal the flaws in uncritical associations of science and democracy of the day and draw our attention to an important pattern: in many places of scientific and technical practice at mid-century, the execution of good science was believed to *require* segregation.

At the very least, for many whites good science precluded the presence of significant numbers of African Americans. Blackness and technological achievement to a large extent remained, even in an era of encroaching civil rights reform, mutually exclusive categories. In engineering, this was true on all levels of technical training and employment, from undergraduate engineering programs through the highest levels of research and development undertaken in academic, corporate, and government sectors. A tiny number of programs, notably at MIT, Howard University, and one or two black land-grant schools, graduated black engineers in this era; in the 1930s, the total appears to have been around a few dozen each year from all programs combined. But what historian Roger Geiger refers to as expanding "social demands for useful knowledge," expressed in the growth of white university graduate and research programs across the sciences at mid-century, did not include a perceived need for knowledge produced or wielded by African Americans. The University of Maryland system embodied many of these beliefs about race and technical competency in the 1940s and 1950s.[6]

If the UMD system under President Byrd's leadership had drastically underfunded its "Negro branch" at Eastern Shore through the 1930s, by the end of the 1940s it appeared on the surface that the campus' fortunes had changed. Byrd, supported by the Maryland legislature, directed unprecedented funding toward the black land-grant school after World War II. New buildings, new faculty, and new programs proliferated as annual legislative appropriations for that campus rose from $33,000 to $113,000 in a single jump in 1947. For the first time in the school's history, it would appear that Eastern Shore was to receive a large enough share of Maryland's public higher education funds to at least begin the task of assuring work of "comparable quality and standard to that at College Park," as Byrd now promised.[7] Byrd foresaw "better paid and larger faculty, more efficient teaching for students, and the beginning of research" at Eastern Shore. But few who had watched Byrd's career and the enduring pattern of Maryland's legislative inaction on black higher education were optimistic. One group of advocates for higher education in Maryland wrote, "The record of Dr. Byrd and the Board of Regents at the University of Maryland in the development of Princess Anne College has been so unwholesome as to give us no confidence in their good intentions toward the higher education of colored youth."[8] Byrd's proposal for "fulfilling the State's obligation to the colored race" struck his critics as paltry: "By no juggling of facts and figures or magician's tricks" would his plan create at Eastern

Shore schools of law, medicine, agriculture, pharmacy, and engineering anywhere close in quality to those offered to white students enrolled at College Park. This was a last-ditch effort to preserve the separate but equal, racially divided system in Maryland's land-grant institutions.[9]

A desperate move from its inception, Byrd's development of Eastern Shore as separate but equal was ultimately a short-lived project. By 1954, the U.S. Supreme Court had ruled that separate but equal education was unlawful, and Maryland faced the end of legal segregation. It no longer had any lawful means of excluding qualified black students from College Park or any other white institution. But the brief period in which Byrd tried to build up facilities at Eastern Shore and forestall desegregation is of historic significance. Watching Byrd flail and reach for arguments that would support the maintenance of UMD's color line, while courts ordered the admission of African American students to increasing numbers of university programs around the country, helps us understand the extent of his commitment to that regressive social system. Nearly twenty years after Donald Murray was admitted by court order to the law school of the University of Maryland in Baltimore, at a time when twenty African American students now studied law there, Byrd was still clinging to the idea that a dual educational system was both optimal and realistic.[10]

The range of parties to whom Byrd promised continued segregation, even as the outlook for *de jure* discrimination clearly worsened, makes clear the varied kinds of political and economic interests of the day that were directly aligned with this retrograde agenda. Politicians, corporate bodies, and white communities throughout the state turned to Byrd as the savior of segregationist values in changing times. But perhaps more pertinent for this inquiry into patterns of discrimination in engineering were the many groups working in opposition to Byrd. These voices advocated in a comprehensive way for desegregation and systematic improvement in government support for black higher education, yet they too were not actually advocating for the end of racial hierarchies in U.S. higher education. As had been the case in the first decade of Byrd's presidency, after 1944 he did not represent a body of thought entirely distinct from that upheld by race reformers. Locating Byrd's strategies and those of his adversaries on the continuum of views about black educational equity makes clear the narrowness of that spectrum.

The UMD system's approach to race in the immediate postwar period reveals a great deal about the enduring racial exclusivity of engineering occupations in the United States. Both the ardent segregationists to whom Byrd's strategies appealed in the later 1940s and early 1950s and such seemingly progressive bodies as the U.S. Office of Education projected a socially disunified world of technical and economic practice. The likelihood of black citizens participating freely and fully in U.S. science and engineering was limited by both political credos, but the two outlooks were not by any means equivalent. The segregationists were often virulent in their opposition to black opportunity and comprehensive in their efforts to keep African Americans out of skilled occupations. Byrd's propitiating rhetoric in support of the separate but equal doctrine, for example, only thinly disguised his distaste for the unhindered entry of blacks into the technical workforce. Federal policy makers, by contrast, acknowledged and condemned some structural causes of black underrepresentation in the skilled workforce. But many of these same proponents of desegregation nonetheless described African Americans as ill-prepared not just educationally, but psychologically and even morally, for scientific careers. In *and* beyond the southern states, beliefs about the innately different occupational capacities of black and white Americans persisted.

World War II and the New American Workforce

As the nation entered the global conflict in 1941, and pressure mounted on colleges and universities to expand the technological workforce, differentials in Byrd's vision for black and white higher education were made newly obvious. Eastern Shore had never offered engineering courses, but as the potential of College Park to develop significant engineering programs became evident, the lack of technical programming at Eastern Shore was more pronounced than ever. During the war, as was the case at many other U.S. universities, UMD's College of Engineering at College Park trained men and women in civil, mechanical, and electrical engineering, and the newly established field of civil aeronautics. Byrd was particularly enthusiastic about widespread programs inaugurated by the Curtiss-Wright Corporation to train women in engineering disciplines of importance to aircraft manufacturers. He welcomed one such program to College Park, and saw it as an initiative that was badly needed and beneficial "for the girls and the country."[11] While extolling the virtues of this newly inclusive gender policy, Byrd continued to resist the idea that black citizens might fulfill new and unlimited roles in the nation's technical spheres. That category of social difference, unlike gender, did not yield in his mind to wartime concerns about the national workforce.

Certainly no black students were to be welcomed to training programs underway at College Park, even in light of the national emergency. At Eastern Shore, training provisions for wartime labor needs were minimal. The campus offered space to wartime programs of the National Youth Administration (NYA), a work relief program developed by Roosevelt at the same time as the Civilian Conservation Corps. Through the Depression era, the NYA trained several million young people in construction, trades, home economics, and clerical areas. NYA participants held research positions as well, assisting university faculty in physics, chemistry, and physiology. At the outset of the war, the NYA established sites for training black and white youth for defense industry needs. Aubrey Williams, head of the NYA from 1935 to 1943, held progressive views on integration and in 1942 he oversaw the creation of an integrated NYA training facility in Harlem, New York, providing skilled machining services for the Brooklyn Navy Yards. Local politicians and officials of the NAACP and National Urban League saw the center as innovative in integrating black and white trainees. It is probably not surprising that at Eastern Shore a different model was in use. NYA students there were not only of a single race, but engaged in such low-skill projects as mattress- and broom-making. And, as local reactions to a fatal fire in the NYA dormitory at Eastern Shore in 1941 recorded, this was a program undertaken at the campus with minimal resources and in dilapidated facilities.[12]

Between 1944 and 1945, a period in which the federal government gave nearly a million dollars for war training and related programs to the UMD system, no additional money beyond the set $15,000 already allotted for Eastern Shore's normal operations went to the black institution.[13] This level of technical contribution to the war effort was not unusual among the HBCUs: when the U.S. Office of Education assessed black educational contributions to the war effort in 1943, Negro colleges were seen as "failing to qualify" for federal programs in the Engineering, Science and Management War Training Initiative.[14] Howard University undertook high-level scientific and engineering research for the Atomic Energy Commission and other patrons, but in this remained an exception among HBCU research units. Byrd's dismissal of Maryland's black citizens as ineligible for wartime technical training meant that Eastern Shore could hardly hope to participate in such efforts. Even during the war, federal influence was unlikely to trump state plans regarding black education. The U.S. Office of Education issued no clear criticism of the states that had failed to equip Negro public colleges for participation in wartime engineering training programs; it is clear that the situation perturbed some federal observers but not to the point of intervention.[15]

Throughout the war years there were few sites around the nation that considered the "manpower shortage" a reason to draw minority citizens into technical training schemes; only a tiny number of African Americans found work in the wartime aircraft industry. Baltimore maintained a "Colored Defense School" that offered training in riveting, industrial sewing, and other skilled trades needed during the war, but not in engineering. The NAACP, on inspection of this school, reported poor facilities, overcrowding, and unsanitary conditions.[16]

When returning white veterans studying under the GI Bill swelled the College Park campus after 1944, Byrd oversaw rapid growth in many academic programs. As University of Maryland historian George Callcott describes, with 8,000 new students enrolling there at the war's end, Byrd enthusiastically marshaled resources for the school's divisions in liberal arts, sciences, business and public administration, home economics, and teacher training. His entrepreneurial achievements in engineering may be among the most dramatic of this period. Overall, the College Park campus had engaged relatively little in wartime research, a pattern that Thelin sees as typical among state universities. In the 1930s and 1940s, the "prototypical American state university" was not, Thelin writes, "first and foremost a home for advanced scholarship." Dramatic wartime research and development contributions most often emerged from the relatively few schools that had longstanding commitments to high-tech research, such as CalTech and MIT.[17]

For Byrd, surveying the research landscape as the war neared its end, this appeared to be a missed opportunity. With characteristic personal force, he persuaded aircraft manufacturer Glenn L. Martin to donate $2.5 million to College Park's engineering programs in 1944. Martin had considered giving this funding to Johns Hopkins, but Byrd prevailed and brought the transformative gift to UMD. The postwar enrollment explosion boosted engineering at College Park from 651 students in 1945 to 1,680 the following year, giving Byrd a newly persuasive case for additional state funding. He was successful in this bid, and by 1949, three classroom and laboratory buildings and the Glenn L. Martin Wind Tunnel brought UMD's College of Engineering one very large step closer to competitive research standing. Fully equipped by 1952, these facilities continued to attract additional private and public funding and expand into emerging fields of specialization, particularly in aeronautics. UMD faculty became associated with significant research in engineering for the first time, and new faculty positions in the College increasingly went to persons holding doctorate degrees.

Martin's decision to endow UMD, rather than Johns Hopkins, gave Byrd not just financial but philosophical bases for the development of engineering at College Park. Martin's chief executive officer explained the company's choice by noting that its own engineering staff had favored Johns Hopkins because that school's basic engineering facilities and its reputation "surpassed those of the University of Maryland." Nonetheless, the CEO noted, Martin himself ultimately rejected the private school and opted for UMD. Martin felt that the principal object of the endowment was to benefit "the industry and public generally" and to do so over the long term. He envisioned the UMD engineering school serving as "the nucleus for the construction and operation of facilities for aviation research and education that will be outstanding nationally and inter-nationally [sic]."[18] A significant number of public figures had tried to promote Hopkins as the more appropriate site for the development of state-funded engineering than UMD; Byrd now put their objections to rest. He quoted Martin's point in testimony to the Maryland legislature, adding, "Now, the State cannot afford to say, in effect, that the above decision was a mistake, and that the gift should have been made to the private institution instead of the State."[19] However, his defensive tone indicates that while Byrd maintained significant power among Maryland's political and corporate elite, his influence was not unlimited. His plans for the UMD system interwove social and instructional agendas and through the battles he faced as system president one can see contours of opposing, progressive movements in the state.

The Unseemly Opposition

As long as segregation lasted, legislative provisions for black higher education in Maryland involved deciding which of the black colleges would receive what proportion of a distinctly limited pool of state or federal money. Through the 1940s, Eastern Shore fulfilled a token role for those running the UMD system and controlling its resources. It survived, but faculty and administrators there were given few opportunities to enact constructive reforms at the school. Increasingly, more liberal groups accused Byrd and the UMD Board of Trustees of deliberately maintaining the school in that token capacity to avoid a genuine effort at improving black higher education in Maryland. Debates about the relative roles of Morgan State College (since 1939 a state-owned school) and Eastern Shore (known informally through these decades as "Princess Anne" after the town in which it stood, and formally as "Maryland State College") also reflect profound tensions about where in the state's economy black citizens might expect to find opportunities.[20]

During the 1930s, publicly appointed commissions, often led by avowed proponents of Morgan College, had found severe shortfalls in the state's provisions for black higher education. The Soper Commission report of 1937, overseen by Judge Morris Soper (a Morgan College trustee), had been particularly scathing about Eastern Shore. By the mid-1940s, the pressure for integration had grown sufficiently strong that the Maryland legislature appointed a new body to study the situation. The chair of the commission was William L. Marbury, and the resulting 1947 "Marbury Report" gave an almost entirely negative account of the Eastern Shore site. As had been an element of such assessments for at least a decade, the Marbury Report proposed that the closure of Eastern Shore be accompanied by the expansion of Morgan State College.[21]

Comparing the Soper and Marbury reports, it appears that ten years and a world war had made little difference in the relative status of Maryland's black and white land-grant schools.[22] The Marbury Commission deemed the state's treatment of the Eastern Shore campus to have been "disgraceful" and "shameful." The campus suffered from low enrollments, an insurmountably remote location, and "deplorable conditions." Byrd believed that the Marbury Commission, like several before it charged with study of Negro educational opportunities in Maryland, was biased against UMD, holding priorities "inimical to certain interests of the University of Maryland." Whether or not those supporting Morgan State College displayed "questionable ethical practices" and "selfish interests," as Byrd maintained, outside observers also found Eastern Shore to be far below standard. Even those who did not want to see the Eastern Shore campus abandoned noted its impassable roads and other shortcomings in its physical plant. Despite the lack of investment in its infrastructure, with its enrollment of only 163 students the campus had extraordinarily high costs per student, and the ongoing argument that Morgan State College be designated as the recipient of federal "Negro land-grant funds" found new advocates. Through such an arrangement, state resources spread across the publicly supported black colleges would be consolidated and duplicative programs and services eliminated.[23]

Byrd resolutely resisted that vision. Instead, he began to lobby state legislators to channel new funds to the UMD system, with the goal of improving the Eastern Shore campus. The opposition solidified to defend the centrality of Morgan State College to public higher education for blacks in the state. A group led by prominent African American businessmen formed the "Citizens Committee on Higher Education" and in 1949 issued a pamphlet comparing the income between 1940 and 1949 of the University of Maryland branches at College Park and in the Baltimore professional schools; at Eastern Shore; and at Morgan State College. Including all sources (state, federal, and student tuition and fees), the two black campuses showed a combined income of around $4 million for the decade, while the white portions of the UMD system together showed more than $50 million. Eastern Shore, having functioned on less than $500,000 for that entire period, represented expenditures of less than 1 percent of the UMD system's overall revenues.[24] One journalist at the *Baltimore Sun,* continuing the paper's tradition of criticizing Byrd, noted that Morgan State College already had creditable science and technical programs that could now be enhanced. To expend resources on those subjects at Eastern Shore, the journalist claimed, where virtually nothing had yet been done, was at best wasteful and at worst disingenuous.[25]

Whether the *Sun* could be accused of issuing "propaganda" or engaging in "unseemly" behavior, as Byrd suggested, it was hardly alone in calling for the cessation of Byrd's improvement of the Eastern Shore campus. The NAACP and other progressive groups based in Baltimore objected strenuously to Byrd's strategies. Adding insult to injury for Byrd, the Marbury Report endorsed the idea that the Johns Hopkins engineering school now become the state engineering agency, an official status that would bring not only legislative appropriations, but also a great deal of paid testing and research work and heightened prestige to Hopkins. This must have been a particularly galling point for Byrd, given that by now he had enlisted Glenn Martin as a major benefactor for UMD's College of Engineering. Byrd's reactions to the Marbury Report were swift and strong. His office issued a statement that arguments for the closure of Eastern Shore arose from "resentment" on the part of Morgan's administrators and alumni. What is more, Byrd said, their case had been aided by "slanted reportorial stories of a large newspaper in Baltimore" that had long been on "the warpath" against UMD's scheme for improvements at Eastern Shore. According to Byrd, their case thus had little merit and could be dismissed.[26]

Byrd's ire at this point may have arisen from a sense of personal affront rather than from deep concern that his new plans for Eastern Shore would fail. After all, the UMD president had his own cadre of powerful backers, including (on most issues) Governor Lane, and many of these figures held the purse strings for higher education in Maryland. Indeed, in 1949 the Maryland legislature allocated $1.2 million to the Eastern Shore school, now to be called Maryland State College in blunt declaration of its public nature. Byrd worked to attract instructors to Princess Anne from southern HBCUs, offering competitive salaries and flexible teaching loads, to establish a larger and more credible faculty. Perhaps most significantly, he hired John T. Williams as the new president of that branch. This appointment continued Byrd's habit of closely controlling the black land-grant's administration, but he now sought that control in ostensible pursuit of parity between the black and white campuses of the land-grant system.

The Ambivalence of President Williams

Williams, who held a PhD in education from the University of Indiana, had previously served as dean at Kentucky State College, another black land-grant school located in a border state. He was to serve as president at Eastern Shore until 1970, not infrequently amid controversy. An unnamed source quoted in the *New York Times* in 1948 described Williams as "an interloper brought here to do the biddings of Dr. Byrd." Many in Maryland saw him as uninterested in fulfilling genuinely ambitious educational goals for the state's minority students.[27] In a number of respects, Williams appeared to have been conciliatory toward Byrd's segregationist schemes prior to 1954; after that date, he was viewed by many as tyrannical in his tendency to ignore student and faculty wishes. This approach matches the classic "plantation style" administration that some historians of higher education have associated with HBCU leadership in this era, and his critics offer a fair amount of evidence for their case against Williams.[28] Yet, to have offered radical alternatives to Maryland's biracial system might have seemed unrealistic to Williams in this period; Williams ultimately spoke of encountering an "invisible wall of racism" and legislative apathy while serving as president at Eastern Shore. It is not obvious that he held insincere or limited ideas about the development of the black land-grant campus at every juncture. While he does appear to have capitulated to conservative social doctrine regarding race, especially in the 1940s and early 1950s when he often supported Byrd's separate but equal strategy, his experiences and ideology nonetheless reflect a complex and often intractable set of obstacles to progressive social action and are worthy of careful examination.[29]

In some instances, Williams simply excused choices made by Byrd that had clearly disadvantaged the black campus. Williams explained that the Eastern Shore campus had remained small and unaccredited through the late 1940s "because of war" (by 1944, the entire enrollment at Eastern Shore had dropped to 55 students).[30] However, as noted above, the war had also brought unprecedented opportunities to white universities and Byrd had embraced these for College Park. Williams routinely deflected blame away from Byrd for Eastern Shore's problems. An unattributed tract of legislative testimony from this period, provided either by Williams or someone else familiar with and favorable toward established operations at Eastern Shore, argued that Byrd had been an advocate for Negro education who had "personally" obtained funding for a new library built on the black campus. The speaker further explained that "we advanced very little from 1890 to 1937," the period in which Morgan College had held direct responsibility for operations at the Princess Anne facility. Once the UMD system took over, moreover, it was not Byrd who then deprived the school of resources, but "the state." Williams, in attributed testimony of the same period, claimed that "no broad program could be developed on such a small amount of money as the state has been appropriating." UMD, Williams offered, by contrast will "bring real respectability to Princess Anne among Negro colleges," thus exculpating Byrd *and* making the case that those who supported a dual educational system in Maryland were unassailable friends of black higher education.[31]

Williams drew on the subject of race selectively to highlight his agreement with Byrd's perspective. In his lengthy statement to the Maryland legislature of around 1949, race is not even distinguished as a topic of discussion until nearly the end, when Williams preceded his thoughts on race relations in the state with the phrase, "Perhaps I shouldn't mention it, but . . . ," as if differentials in black and white higher educational provisions were not the very subject of the congressional hearings! Other frankly conciliatory statements by Williams directly bolstered Byrd's idea that the end of segregation would cause untold social disruption in Maryland. Defending the maintenance of the land-grant facility at Eastern Shore, rather than the transfer of federal funds to Morgan State College, Williams noted that, "White people in [the town of] Princess Anne want us . . . those in Baltimore don't seem to."[32] Williams apparently addressed an ambient fear of racial unrest when he informed the legislators, on whom he was dependent for funding, that "we have no wish to become a discordant element" and, even more self-deprecatingly, "all we ask or shall ever ask is that you deal fairly with us, and we have confidence that you will."[33]

Byrd himself routinely discouraged black activism, warning the NAACP in 1948 that by "pushing too hard" and thus aggravating a "delicate situation" it might lose out on "greater opportunities than even were dreamed of in previous years."[34] In making points like these Williams must have presented a reassuring figure to Byrd, especially in comparison to the many activists now demanding full-blown reform of Maryland's land-grant system.

But a distinct ambiguity infects Williams's next point in this same passage of testimony: "The Negro people of Maryland are never going to get anything more than the white people want us to have."[35] Read in proximity to the preceding assurances, those words convey a certain compliance among black Marylanders; as if nothing more than is willingly given to them by the white people of Maryland shall be demanded. However, read independently, a sense of dismay and even bitterness comes through in these words. Williams may have been subtly signaling the "wall of racism" to which he referred years later, near the end of his tenure, and the truly retrograde climate in which his school was trying to offer meaningful higher education to black citizens. Writing in the late 1950s, Williams indicated that many of his efforts to cultivate support for the college in the local community had "died a-borning." Countering his one-time claim that blacks were made welcome by the white community of Princess Anne, he now wrote, "It is no mere conjecture to say that the principal reason that the local business and professional men have not come to the fore in support of the college is ethnologic."[36] This seems a more believable portrayal of race relations around the Eastern Shore campus than the earlier one; the town of Princess Anne experienced racial confrontations through the 1960s as local white business owners resisted integration. In 1966 one journalist described the town as one which had long held a state college but "never became a college town." However upbeat and conciliatory his language to state officials may have been, Williams likely had moments of profound frustration in the daily operations of the black campus.[37]

Williams's skill and wavering levels of sincerity in promoting black higher education at Eastern Shore could certainly be viewed critically; in whose interests was he operating when he defended Byrd's retrogressive racial strategies? But his apparent ideological inconsistencies may also be seen as pragmatic responses to white authorities on whom his school depended for its continued existence. Had he pursued a more adversarial course with Byrd, he might well have lost his job. More significantly from a historical perspective, Williams's rhetoric before 1954, especially at its most accommodating, casts a raking light on the duplicity of separate but equal policy makers. When Byrd and other segregationists claimed that Eastern Shore, not Morgan State College, deserved to receive federal land-grant funds, they hoped to contain the educational opportunities and attainments of blacks while still deflecting criticism coming from desegregation proponents. Morgan State College was indisputably the larger and more fully developed institution in this era. But very few of the state officials in charge of higher education, even among Byrd's opponents, envisioned a black citizenry with full economic equality and it is unlikely that Williams would have found support among Maryland officials and lawmakers for truly significant expansion at Eastern Shore. State officials such as Marbury who advocated for Morgan State College undoubtedly held more progressive hopes for black public higher education than did Byrd and his backers, including Williams, but for the most part they did not present a radically remade racial ideology.

For all its reformist intentions recommending proportionate funding, nondiscriminatory salary structures, and the pursuit of accreditation at sites of black higher education in Maryland, the Marbury Report did not endorse desegregation for the UMD system. Only the "Minority Report" of the Commission called for an end to the dual system, and that only at the graduate and professional level, not for undergraduate programs. The majority report sought separate but equal provisions, to be enacted through increased funding for Morgan State College and for the State Teachers College at Bowie. It also recommended, "that specific provision be made for the land-grant college for Negroes to have an equitable share of the federal funds allocated to the state for agricultural extension and for research in agricultural and mechanical arts."[38]

There is a telling omission in this recommendation: unlike most white and integrated land-grant universities in the United States, Maryland's public university for African Americans, wherever it was ultimately located, would not embrace engineering research or advanced instruction in technical disciplines. Conventional associations of blacks with rural lifestyles continued, justifying a raft of assumptions about blacks' lack of fitness for participation in the nation's industrial spheres. In these associations, the segregationist interests of Maryland could find support from federal quarters.

The Racial Ideologies of Federal Education Policies

Postwar ambitions for the technical education of white students at College Park, where the UMD College of Engineering now prospered, were nowhere to be heard at Eastern Shore. Even relatively progressive planners for the economic uplift of black Marylanders, who hoped to erode Byrd's discriminatory agenda, implied a ceiling on black occupational attainment. The federal government did not transcend this view in any consistent way. While President Roosevelt embedded some racial reform projects in New Deal legislation, and some educational and employment gains occurred, a close appraisal of the government's approach to black higher education reveals distinctly regressive features. Policy makers' depictions of African Americans as outside the mainstream of most professional sectors, including engineering and other branches of the sciences, seem too extensive and thoroughly conceived to have arisen merely as gestures of accommodation to segregated southern states. They have a complexity and internal logic that suggests deeply held convictions about racial difference.

The U.S. Office of Education unit responsible for "Improvement of Agricultural Education in Negro Land Grant Colleges" in this period argued, perhaps predictably, for the benefits that agricultural careers would bring to individual black Marylanders, to all blacks within the state, and to the state as whole. After all, this department was created to promote agricultural programs at black land-grants; one could hardly expect it to denigrate agricultural livelihoods. But the nature of this unit's work was based on discriminatory racial outlooks. No parallel units for improvement of engineering education or business education at black colleges existed, despite the growth of those fields at predominantly white land-grant schools. As in the early and mid-1930s, when black labor flowed northward and from rural to urban settings, rhetoric now offered by the Office of Education (absorbed at the start of the war into the Federal Security Agency alongside Social Security, and Food and Drug programs) projected a sort of rural colonialization. One report written expressly about Maryland in 1948 by unit director R. M. Stewart proposed that Eastern Shore undertake careful curricular reform, better record keeping, and enhanced outreach to the black farmers and housewives of rural Maryland. The campus might thus become a center of agricultural immersion for the twelve coastal counties of Maryland, and increase its enrollment from all parts of the state. This growth would lead to expanded land ownership among the state's black citizens, facilitated by the preparation of well-trained farmers through degree and extension work at the land-grant school. The goal was to improve the lot of some underemployed black Marylanders, yet the report clearly recommended maximizing the number of black citizens engaged in agricultural work in lieu of other occupations.[39]

Stewart held that "the college enrollment must be quadrupled immediately" and that "definitely ... there is relatively slight representation of [Maryland's] Negro population in the College." Agriculture, after all, provided a field in which, "the humblest of men can gain equities in the business by intelligent and frugal living." This was a "people's college," and it would best fulfill that function by bringing opportunity of this kind, providing a "more secure and satisfying life than can be attained elsewhere."[40] Significantly, the creation of more numerous and more stable farming operations among black Marylanders had another purpose in Stewart's eyes. It would serve a segregationist function achieved through two related factors: advocacy for the maintenance of two racially distinct economies in the state and the promotion of a new black "leadership" to be drawn from among educated black men. In both respects Stewart represented a wider ideology within the federal government that reconciled progressive reform in higher education for African Americans with distinct limits to black participation in science and engineering sectors.

In the first instance, Stewart helped make the case that Negro farmers would function in a separate economic sphere from white agriculture. Among other things, he pointed out (as had advocates for black agricultural training in Maryland in the previous decade) that good nutrition would bring about improved standards of living for rural blacks, and that "the proportion of Negro farmers in Maryland should be enlarged to meet the need for improved nutritional standards." He signals two important concepts here. One is that due to this debt of service that people of color owe their communities, African American businesses should seek to supply African American rather than white consumers. If followed, that construction of ethnic obligation would discourage black participation in white markets. Second is the idea that blacks, and only blacks, should fulfill blacks' market needs. Whites can neither be asked nor expected to do so. In other words, Stewart instructs his readers that unless a source of goods and services arises from within the black community, no other services will be available there. Stewart was quite explicit that through an expansion of agricultural programs at Eastern Shore, "young men should be encouraged [to meet] improved nutritional needs of the Negro population." This logic wedded UMD's black branch to the provision of "a sufficient supply of food products" for the state's black citizens and also optimized the economic isolation of blacks.[41]

Byrd responded warmly to Stewart's recommendations, affirming to Stewart that these suggestions would help Byrd and Williams make the school "something we can all be proud of."[42] Conversely, Byrd roundly criticized Fred J. Kelly, employed by the Office of Education as "Specialist for the Land Grant Colleges and Universities," for Kelly's flexibility regarding the location and curricular role of land-grants. In 1949, Kelly, one participant in the extensive 1943 and 1948 Office of Education surveys on Negro higher education, supplied Morgan State College proponents with data asserting the value of a nonagricultural public college for Maryland's Negroes, ideally to be located in Baltimore. In answer to Kelly's apparent support of Morgan State College, Byrd fumed, "Frankly, I could not conceive of a man of your experience among the Land Grant colleges making a statement in which agriculture is so minimized."[43]

It seems highly unlikely that Byrd maintained that position out of some unalloyed sense that Negro citizens of Maryland would maximize their life chances by attending Eastern Shore, rather than Morgan. Unattributed contemporary legislative testimony against the abandonment of Eastern Shore, probably supplied by Byrd or his office, clearly marks agricultural training in this era as a means of suppressing black activism. In passages supporting the development of the rural campus, the reader gets the sense that to keep blacks out of the city was to forestall social difficulties: "a location distant from an urban center eliminates the distractions of city life, gives students better opportunities to study, develops among them a greater community spirit, and eliminates many of the pitfalls to which students in college are subject.... The movement from the country to the cities already offers one of our greatest national problems and we are trying to find ways to combat that."[44] As the primary site of racial unrest in the United States from World War I onward, cities were central objects of concern for both the left and right, as subsequent chapters will show. The universalizing language used here, denying the cultural advantages of urban higher education, aligns with the most conservative doctrine of the segregationists.

Stewart projected a world in which Byrd's orderly, segregated vision for Maryland in the postwar years could be one with advantages for all citizens, black and white. Separateness carried no intimation of disadvantage. The U.S. Office of Education, in its comprehensive set of recommendations about black higher education issued in 1948, carried this logic beyond agriculture to encompass segregated medical, legal, and other professions: "A qualified Negro is more likely than anyone else to have the knowledge and sympathetic understanding of the Negro's background and present conditions."[45] Here, separate spheres of professional practice or economic activity are construed not only as fair, but as desirable from the perspective of the services to be offered. Not least remarkable is the idea that the free enterprise system works even when fractured into distinct markets—the distribution of economic opportunities to all parties in the system was apparently not impeded by this race-based disunification. It is perhaps not surprising that when the Office's report outlined what opportunities awaited blacks in science and technology, it implied a parallel disunity. Unlike policy makers in Maryland, the federal officials behind this report projected the entry of some minority citizens at all levels of technical occupations, including as civil engineers. Blacks, once trained in engineering, could look forward to employment on infrastructural projects related to "housing and slum clearance" and on expanding public works and public service projects in "Liberia, South America, Virgin Islands, Ethiopia and certain parts of Asia and Europe." Modernizing knowledge could be deployed in separate spheres on the global as well as national context, and an expanded pool of practitioners of different racial or ethnic identities would bring this about.[46]

Simply in allowing for black entry into professional engineering, the federal government was evincing a more progressive vision than did Byrd, but how much more progressive? How the operation of separate technical and economic spheres could possibly have been construed as fair is a vexing question. Where merit ostensibly serves the function of sorting commendable technical practice from fraudulent or flawed practice, as in engineering, surely unhindered competition is needed to assure true comparisons between practices. The ideal of objective standards for applied knowledge would seem to demand an even playing field on which the "best" person might win. The innate subjectivity of such judgments aside, dividing the world of technical thought and practice into separate spheres of operation surely voided that proposition.

Materials issued by the Office of Education on black higher education in this period are at times difficult to interpret. Many were produced under the direction of Ambrose Caliver, a complex character who served the agency from 1930 until 1972. Caliver, one of very few African Americans holding authoritative federal positions at the start of this period, has been thoroughly described by Peyton Hutchison in his aptly titled doctoral dissertation, "Marginal Man with a Marginal Mission." Caliver held a doctorate in education from Columbia University and developed a specialization in the education of minority groups, adults, and the aged. First enlisted by President Hoover to study black opportunities in higher education, Caliver for decades undertook the difficult task of representing minority interests to ambivalent patrons. His devotion to this work is indisputable, and Hutchison conveys Caliver's imperturbability and profound empathy for all disadvantaged persons. But throughout his professional activities, Caliver also had to accommodate majority resistance to the equitable preparation of blacks for professional practice in science and engineering fields. Sensitive to the risks of overreaching in his recommendations for reform and courting controversy, Caliver seems often to have turned to projections of black opportunity that would only partly challenge prevailing white expectations. As Hutchison sees it, Caliver formulated "strategies of value to members of minorities who seek legitimate goals within established bureaucratic organizations." Some Office of Education publications written by Caliver in the 1940s carried intimations of a disciplinary function for black technical education, drawing an oppressed minority into somewhat compliant participation in the labor market despite dismal prospects of encountering authentic opportunity there. One idea in particular that highlights this intention in the Office's reports and other documents on Negro education is the creation, through education, of black "leadership" for the nation's African American communities.[47]

Leadership and Social Order

In discussions of race in the pre-*Brown* era, *leadership* was a term laden with meaning. Its usage at mid-century by black and white policy makers varied considerably from W. E. B. Du Bois's notion of a "talented tenth," first conceived by him in the 1890s. Du Bois implied that a cohort of exceptionally qualified college-educated African Americans would lead the processes of political reform in the United States necessary for the achievement of racial equity. Some scholars recognize an elitism behind Du Bois's coinage, but generally agree that the phrase does not imply race separatism.[48] By contrast, in Caliver's 1943 report on Negro higher education, the U.S. Office of Education equated black leadership with "the ability of a region to solve its own problems," intimating if not directly calling for a racial separatism in U.S. society (although not in the militant sense that ideologies of black separatism later emphasized). Furthermore, that "ability" required a "broad base of effective elementary and secondary education and a program of higher education."[49] This report cannot be construed as a federal call for equal educational opportunity for black citizens, let alone for the end of segregation. Rather, the cultivation of black leadership projected a geographic and occupational distribution of citizens on the basis of color. It was from the ranks of Eastern Shore graduates, Stewart noted in his survey of black colleges in Maryland, that the state's black population might draw its leadership, gaining much from the salutatory "influence of the educated negro."[50] And this was to be a particular kind of education, in particular disciplines. Stewart, like the writers of legislative testimony supporting the maintenance of the black land-grant at Eastern Shore, associated black achievement and social leadership with rural life. In refutation of arguments that Morgan State College be designated as the recipient of federal funds for Negro colleges, and the Eastern Shore campus abandoned, a UMD spokesperson offered this view of black higher education in Maryland: "We need most to have a kind and quality of education which will develop for us the right kind of leaders, men and women who will keep their feet on the ground and lead their people in ways of usefulness and harmony. And gentlemen, the backbone of our way of life came from the farms, and the nucleus of it is still there."[51] This vision would find support in the federal position on black higher education during the 1940s.

The Office of Education's 1948 bulletin on Negro life and education in the nation, authored by Caliver, notes an "ever-growing demand for leaders in all walks of life," which makes imperative financial support for more graduate and professional study among African Americans. The trope is repeated in nearly every document assessing black educational and economic opportunity in this era.[52] Not unlike rhetoric that described white college graduates assuming leadership roles in economic or scientific spheres, common in treatises about U.S. higher education earlier in the century, these invocations of black leadership carried implications for social structures well beyond the university. After about 1890, that earlier educational advice had helped elevate white, male citizens to positions of professional attainment; its tenets linked academic achievement and social authority. But unlike those constructions, which boosted the cultural standing of educated white men, notions of black leadership implied constraint of a potentially disruptive faction of the American workforce.[53] On one level, some policy makers and analysts, both white and black, believed that the black population tended toward socially problematic behavior that had consequences for the race as a whole. This tendency was supposedly manifest in the general black population through poor hygiene habits, inadequate parenting, or rowdy and disrespectful behavior, as Chapman put it in his overview of black higher education in the 1930s. On another level, these social transgressions could also be associated more subtly with insubordination that could lead to race activism and unrest. In either case, analysts linked their subjects' public and private conduct. To counteract this transgressive tendency, in theory, a select few persons of color inclined toward commendable social and personal behavior could become models for the less disciplined many.[54]

In educational spheres, advocates of a new, educated black leadership seemed to believe that blacks ordinarily failed to make a priority of intellectual values and instead resorted to easy distractions of social life, an idea with particular impact for the racial diversification of science and engineering. Beyond the patronizing tone of such claims, there is constructed an innate inability on the part of blacks to undertake science and technology work in any effective way. Exceptions, in the form of "certain individuals" of color who had contributed to "the advancement of knowledge" (and are too little documented by textbook writers and publishers, according to the Office of Education's 1948 bulletin), were perceived to indicate the potential of the race. If the accomplishments of these Negroes were more widely researched and published, their example would "serve as incentives to further endeavor" and induce a "greater 'sense of belonging' than many Negroes now have."[55] But the evocation of black potential is itself ideologically loaded here. It is associated with individual capacity and agency, rather than with the structural conditions under which blacks obtain education. Even though the point was made in this period that "environment" or "lack of opportunity" had caused a lack of black achievement or "leadership," particularly in the intellectual or academic arena, blame was often placed on black individuals for their lack of attainment. For example, General Omar Bradley, looking back on World War II, summarized that U.S. blacks had had "good opportunities" and "improved health and finance" during the war, and had become "more articulate" in political matters. Yet, Bradley added, blacks had shown "no leadership qualities." The 1948 Office of Education bulletin listed "racial factors having a deterrent effect" on the creation of black leadership; these included conditions "largely under the control of Negroes themselves" having to do with campus life, family life, and "cultural level." In each of these instances, Caliver identified some underlying individual impediment, attributable to race, mooting the role of environmental or structural factors in black conduct and attainment.[56]

Assessments of black higher education arising from federal bodies like the Office of Education could never be mistaken for the undiluted segregationist educational philosophies of someone like Byrd, but nonetheless they offered a combination of reformist and rearguard ideology. On one hand, as Chapman had found to be the case at the end of the 1930s, by the mid-1940s the Office of Education also reported disapprovingly that in none of the separate but equal colleges or universities for African Americans was there actually any real "duplication" of white programs in agriculture, let alone in "graduate and professional work in engineering and architecture." Nor was significant curricular duplication to be found in "commerce and the trades." What is more, the Office of Education emphasized that merit must be the basis of education "in a democracy" because no other basis fulfilled U.S. social ideologies.[57] On the other hand, in many of the Office of Education studies, Caliver made the case that blacks display a constitutional resistance to the innate logic of science. Again, bearing in mind Caliver's larger goals and the challenges facing all race reformers in these years, we may understand these points to reproduce conventional notions of difference rather than as concerted attempts to construct biologically based racial hierarchies. Yet, whatever his reasons for its inclusion, Caliver's invocation of behavioral bases for intellectual attainment resonates with descriptions of colonial settings, from the eighteenth century forward, in which Europeans and white North Americans detected an inability on the part of indigenous peoples to appreciate logic, efficiency, and the conceptual bases of science and technology. In 1948, the Office of Education saw blacks as lacking a "zest of discovery," and those black faculty who were inclined toward stimulating inquiry were often discouraged by a lack of "intellectually alert" colleagues.[58]

In black colleges, where administrators commonly failed, according to the Office of Education publications, to undertake long-range and institutional planning, graduate and professional study were undermined. This brought about an unsurprising lack of appreciation for the "place of research" in the academy. That most black colleges were severely underfunded and understaffed should have been no impediment to the development of research programs, apparently: "While it is conceded that the total budget for the average [black] institution is inadequate, administrators usually have sufficient latitude to make larger allocations for research than those they now make."[59]

Caliver identified a tendency among administrators of black colleges to discourage the exploration of new fields: "Sometimes this lack of encouragement comes from the administrators' feeling that it is not politic for teachers and research workers—especially in certain fields—to advance too near the frontiers of truth."[60] As seems to have been a perpetual feature of official analysis of black higher education for much of the twentieth century, difficult choices made by black educators in the face of limited resources were cast as "errors" arising from those educators' personal failings.[61]

Some of these problems of academic management, as Caliver's 1943 report briefly recognized, also occurred in small white colleges, but the author nonetheless raised issues of character associated with black identity. Throughout the 1943 and 1948 publications, institutional conditions at the HBCUs that were unfavorable to high-level academic performance by faculty, such as heavy teaching loads, low job security at schools where salaries and tenure were not adequate, and poorly equipped libraries and laboratories, were conflated with attitudinal failings on the part of black faculty. Administrative and managerial shortcomings were seen to either produce or encourage the personality traits of blacks that impeded intellectual productivity. These traits included the lack of an ability to plan, of self-discipline, and of "maturity." Although Caliver acknowledged that these traits may have arisen because "the minority group status of Negroes has given them little opportunity . . . to develop many of the qualities which are essential in research," in the aggregate, he indicated that where uncultivated, the Negro might regress to an unfortunate set of behaviors. The "high plane" on which research must be undertaken was not the level to which the black intellect defaulted.[62]

Science and Compliance

Negative judgments about black culture and support for the validity of racial categories were embedded far below the surface in the Office of Education reports, amid lengthy recommendations for the correction of race-based educational inequities. The internal contradictions here are striking. The agency supported extensive study of black higher education and published countless summaries of its findings. As Peyton Hutchison makes abundantly clear, Caliver (and perhaps his superiors within the agency) recognized daunting obstacles to inclusive educational policies and sought incremental reforms. Still, in these publications, cultural proclivities were associated with race so that a reader intent on maintaining the status quo might conclude that no improvement of institutional or structural conditions actually seemed likely to remedy black economic exclusion. When Caliver's 1948 bulletin noted that among black communities, "unintelligent and blind acceptance of unsound interpretations of the Bible" often "prevented the use of scientific methods in combating disease or in advancing knowledge," he echoed Chapman's criticism of rural blacks' reliance on home remedies and irrational fear of the medical profession. Both critiques posed black religiosity as an impediment to intellectual achievement. Similarly, Caliver pointed out a susceptibility among blacks to "quackery" in matters of biology or physiology, and indicated that college-educated black community members might become "teachers of hygiene" to offset that weakness. In some ways, this kind of belief may have offered as great an obstacle to black entry into the sciences as Byrd's reductionist justification for the absence of an engineering program at Eastern Shore in 1947. Byrd said only that few blacks wished to enter engineering; the Office of Education provided a detailed basis for that disinclination.[63]

The Office of Education analyses projected a no-win situation for aspiring black engineers and scientists, in which essential personality traits were likely to draw even the well-intentioned black researcher toward improvident and unmotivated behavior. On the one hand, blacks' "lack of economic security" could prevent intellectual participation because material deprivation engenders "fear, timidity, caution and inhibition which are deterrents to the adventurous, expansive spirit needed in research and professional growth."[64] On the other hand, if this formulation implies that increased educational resources might improve black attitudes about science, the likelihood of any individual black person actually manifesting laudatory intellectual behavior was slim. Caliver wrote that when blacks did achieve "social mobility," they were often drawn to the attractions of "sensual" pursuits that could be "the undoing of modern civilization." That statistics about the incidence of syphilis among African Americans also appeared in Caliver's 1943 report, on a list of wartime problems "for which the [Negro] colleges have certain responsibilities," may tell us just how probable such problematic behavior seemed to the author. Caliver may have provided this information as an inducement to improved government provisions for black health care, but it introduces the specter of behavioral deficiencies on the part of African Americans (reiterated by another item on this list that identified black "indifference" to wartime technical training opportunities).[65] If such tendencies were believed to exist among blacks, how compelling was the case for educational interventions that might carry the black thinker to a higher plane? When we consider the Office of Education's guarded outlook for black occupational attainment, Byrd, with his devotion to the segregated world of southern education, does not seem nearly as much of a philosophical outlier as he might.

Caliver, speaking for the Office of Education, apportioned blame for blacks' lack of social and cultural development with care, placing the "larger responsibility" for this regrettable situation on public officials and members of the public who stood in the way of improved support for Negro colleges. Yet, Caliver nonetheless pinpointed the role played by African Americans in their own disadvantage with one notably double-edged argument. The failure of black colleges to participate in wartime science, engineering, and management training programs run by the federal government, Caliver wrote, arose in part from a historic pattern of exclusion that had deprived Negro schools of resources, but also from "confused thinking on fundamental issues" among black educators and scholars. This confusion resulted, Caliver asserted, only "partially" from a lack of training and guidance for Negro college administrators concerning problems and issues "of public policies and human relationships."[66] The other source of this confusion was apparently integral to black educators, not a result of structural inequities. The invocation of "confusion," crucially, signals error on the part of the Negro educators, rather than authentic dissatisfaction with the status quo. Blacks may display agency, but it will lead to misguided action unless they correct their current course. Radical realignments of resources and opportunities in public higher education were taken off the table here. Instead, black colleges were asked to strive harmoniously to integrate their students and faculty into a stratified economic system. The 1943 report declared that the "base of occupational choices [for Negroes] be widened" to include emerging areas such as aviation, radio, and agrobiology. But the very next point is that Negroes also need to cultivate a "higher respect" for the social value of the "simpler occupations . . . in which the majority of Negroes are now employed."[67] Paired with advice to black colleges that they strive for more open admissions and enact social interventions through adult education, such recommendations did not bode well for either institutional development or individual opportunity.

Integration Begins

Against this backdrop of ambivalent reform, intended to correct generations of racial inequities but only to a certain degree, governmental bodies in the United States were also beginning to see the possibility of an end to legal segregation. It is clear that to most of those responsible for public education outside of the segregated states, this would have been far preferable to solutions outlined under separate but equal provisions. The economic and ethical obstacles to creating separate black and white facilities were obvious, even if most progressive voices were not envisioning a truly open society when they advocated for black betterment. We will eventually see that such mixed ideals pervade higher education for minority Americans even today. But amid the growing advocacy for an end to segregation in these last few years before *Brown,* Byrd found himself increasingly marginalized as he clung to separate but equal provisions for the UMD system.

By 1947, with the support of the NAACP or simply inspired by that organization's successes, black students were beginning to apply to UMD graduate programs in increasing numbers, aware that court cases were setting precedents for their admission to other white universities. One African American student, Wilmore B. Leonard, applied for graduate work in chemistry at College Park in 1947. He was offered provisional enrollment, but when all applications for the year had been received, Leonard and at least six other black candidates who sought admission to the graduate school were turned away. Edgar F. Long, head of admissions at UMD, actually traveled to Leonard's home in an effort to force the student to turn over the printed card that granted him provisional admission to College Park. Leonard refused to relinquish the card, at which point Long changed tactics. He told Leonard to keep the card "as a souvenir" and that Leonard's admission had been "a mistake." Long then offered Leonard a scholarship for out-of-state study. Long apparently feared that Leonard might pursue his case in court, or seek media support from the relatively liberal Baltimore press, neither of which appears to have happened.[68]

Byrd began, in this period, to handle all applications from "colored students" himself, rather than having the UMD staff or deans do so. He found himself in a messy situation. Regarding encroaching integration, he fought for "gradual segregation," whatever that might have meant to him (this was an approach that would entail, Byrd said, a "minimum of friction").[69] He was often inconsistent in his claims as he tried to align his audiences. In 1948, he told Maryland legislators that "Negroes have got to get University of Maryland diplomas," but that it would be "utterly impossible for a Negro institution to develop work equivalent to or equal in any sense to the work of the University of Maryland." In the same meeting, Byrd maintained that it would "disastrous to have them [black students] come into College Park under present conditions."[70] Clearly, he had difficulty imagining himself overseeing the transformation of College Park into an integrated university.

In 1949, the Southern Conference Educational Fund, a civil rights organization based in New Orleans, asked Byrd to report on the school's experiences with integration. With black students attending its law school, Maryland was now one of six southern or border states admitting African Americans to one or more graduate programs, and the group was interested in knowing how this policy was "working out in actual practice." They sought to publish an article on Negro graduate work in the UMD system. Byrd replied that "no good would come of publishing such an article as you suggest," and went on to express his unease: "I have an exceedingly difficult problem on my hands, but believe that within the next year or so we will solve it . . . although probably not entirely as either the white group or the Negro group would wish."[71] In the summer of 1948, the Baltimore press excoriated Byrd for trying to operate a separate summer school for Negro teachers. Only two students had enrolled, leading to strong criticisms of wasted public resources. Byrd was apparently beginning to see that compromise was not likely to leave his popularity intact.[72]

Whatever Byrd may have had in mind as a solution for his "exceedingly difficult problem" (and what that solution was is by no means clear from his actions or writings in this period), events overtook him. In 1951, College Park admitted its first African American student in engineering. Drawing on the *Gaines* case involving the University of Missouri, a 1950 decision by a Maryland Court of Appeals had set the stage for the admission of a black student, Esther McCready, to UMD's nursing program in Baltimore. Based explicitly on the fact that Maryland offered no state-supported school in which blacks might study nursing, the *McCready* decision gave strength to another suit already underway involving the exclusion of blacks from engineering at College Park. Hiram Whittle, an undergraduate student in his third year at Morgan State College, had been hoping to train as an electrical engineer when the NAACP asked him if he would be a test case for black admission to UMD's engineering college. The organization filed a suit on Whittle's behalf in Baltimore City Court in 1949. Lawyers for the case included Donald Murray, who had broken ground as the first black student of law in the UMD system. Murray now worked alongside Thurgood Marshall to challenge ongoing segregation at College Park. Perhaps daunted by this experienced and prominent counsel, the UMD Board of Regents held a special meeting at which, following the advice of the state's attorney general, they decided to admit Whittle to College Park "immediately" rather than risk losing the case. Byrd had hoped that Whittle would start his engineering training at Eastern Shore, although it seems unlikely that a curriculum suited to this purpose existed there at the time. However, the attorney general recognized the NAACP as a potentially perpetual irritant, and "hoped to take the wind out of its sails" by admitting Whittle to College Park. The Board of Regents complied.[73]

It seems clear that such decisions made by UMD's leaders about the admission of black students arose from outside pressures, not from a deeply felt conversion to the cause of racial equity. In April 1951, the UMD Board of Regents determined that in order to comply with rulings of the Supreme Court, the professional schools in Baltimore must begin to consider applications from "all citizens of the State, without regard to race, color or creed." Byrd enacted this dramatic reform, writing with uncharacteristic brevity to the dean of the School of Medicine simply that, "It will be in order, therefore, for you to follow this policy."[74] A few months later, Hiram Whittle, preparing to attend engineering classes at College Park, applied for a dormitory room. For Byrd and the Board of Regents, a disturbing situation came to a head. Distressingly, Whittle's admission now involved his "living with white students, attending social events, and so on."[75] The mixing of black and white students had arrived and was, to UMD's leadership, in every respect regrettable.

As the Board of Regents reported to the Maryland legislature in 1951, reflecting on Whittle's admittance into the College Park social scene, "This action would not have been necessary if the previous advice of the Board of Regents had been followed."[76] Ironically, at the time and again many years later, Whittle recalled having encountered little social difficulty upon arrival at the white campus. He attended few social and athletic events, but attributed this to the heavy course load required in engineering. Whittle recalled that the feeling among students at College Park "was fine." Many were from Baltimore, he said, and "used to mixing." The College Park campus "didn't really have a racial problem . . . didn't feel any different from studying at Morgan."[77] The disjunction between Byrd's palpable anxiety about integration and the experiences of this cohort of students is striking.

The UMD Board of Regents was profoundly uncomfortable with Whittle's presence at College Park and expressed a sense of aggrieved stewardship. In 1951, an unnamed spokesperson for the Board of Regents (possibly Byrd himself, given his self-identification at one point in the document) claimed in a transcript of legislative testimony that: "The University of Maryland is part of the public system of education in Maryland that has always been operated on a bi-racial basis. Every act of past Legislatures indicates that the University is expected to carry on under this system."[78] And yet, the speaker continued, separate but equal college facilities had never really been supported by legislative funds, leaving "makeshift policies" in place that had led directly to the admission of Whittle at College Park: "The facts show that the Board, over many years, has made repeated requests of State authorities for adequate funds to meet this need. If these funds had been granted, this action of the Board today would not have been necessary."[79]

At this point, the spokesperson recommended, the best possible option would have been to split programs of study between Morgan State and Eastern Shore, so that between the two schools a set of offerings for Negroes, comparable to those offered to whites at College Park and the Baltimore professional schools, might have been developed. If this could not be achieved, it was suggested, some compromises could be considered. Among these, the UMD system might, "provide for two years of Engineering work for Negroes at Princess Anne, with the two remaining years, and with graduate studies, at College Park." Plans to integrate research at Eastern Shore and College Park had been mentioned before, and now seemed to offer some hope of keeping the black campus open and thus keeping the majority of black students away from the white campus.[80]

In 1951, UMD formalized its plan to admit Negroes to the College of Engineering at College Park, but required that their first two years be spent at Eastern Shore. White students majoring in engineering received their first two years of coursework at College Park, so there can be no doubt that any needed curriculum was available on the white campus. It is evident that without such regulations, the need for Eastern Shore would have evaporated, in which case the trend toward court-ordered integration at College Park might have proceeded unimpeded, bringing unwanted change to the University of Maryland. Byrd, or someone of like mind, in unattributed testimony to the Maryland legislature accused those responsible for the affairs of Morgan State College of wishing to see Eastern Shore "abolished, or emasculated to virtual impotency." But this was not merely a territorial dispute. The testimony purported that the option of opening white programs at UMD "for all citizens alike, without any attempt at segregation regardless of race, color and creed," would be objectionable to "the great majority of the people of the State . . . because of their long-standing custom of bi-racial education."[81] Whether that surmise was true or not, to invoke the notion that the white public had been betrayed gave Byrd and his allies a powerful sense of self-justification.

Conclusions

Clearly, only through a thorough reimagining of the role of race in society would educational policy in Maryland change. By 1952, with *Brown* already working its way upward through the courts, officials were finally conceptualizing the Morgan State College, Eastern Shore, and white programs of the UMD system as constituting a single, racially integrated, educational apparatus. The UMD Board of Regents was moving toward a clear rejection of the dual education system, claiming that however well funded, the continuance of a biracial system would not bring Maryland "three institutions that would be as strong, in any sense, as the one institution open to all." But their arguments for this change were largely fiscal, and run through with blame for a legislature that had failed to deal with a regrettable social situation. In their view, integration was by no means the optimal outcome of the decade's conflicts: "What has been done heretofore neither gives the Negro what he is entitled to nor prevents him entering the University of Maryland. The whole problem is one which involves constant adjustment."[82] With Murray's enrollment in the School of Law, UMD was the "first state university in the South during the twentieth century to accept Negroes" and with Hiram Whittle's admission, "the first to accept Negro undergraduates," as Callcott writes in his 1966 history of the university system; these points are true technically, but do not signal an ideological conversion. The admission of Whittle was believed by the system's directors at the time to constitute a failure, not the attainment of some democratic watershed. It would be decades before UMD's College of Engineering turned to the project of minority inclusion with reliable support from the university system. Its considerable accomplishments in this regard date from the 1970s, when it turned resolutely from those earlier racialized approaches. Today, the college is home to some exceptionally innovative diversity staff and the College Park campus boasts many well-supported multicultural programs. In contrast, Eastern Shore remained an underfunded and marginalized branch of the land-grant system for nearly forty more years. Its standing today is seen by some as representing a very recent turn by the state to genuine concern about the historically black campus.[83]

The case of UMD as it operated through the long civil rights era shows that "southern exceptionalism," as Wayne Urban and other historians have indicated, breaks down when the roles of segregationist and integrationist educators are examined. Despite their many invocations of talent and merit as metrics applicable to blacks, as means of drawing the long-excluded minority into new fields of intellectual endeavors, federal policy makers during this period helped sustain a racial hierarchy in American occupational structures. The U.S. Office of Education, eager in the 1940s to address historic racial inequities and yet, it appears, also complicit in preserving long-standing ideas about racial difference, projected a racial character onto the sciences and other fields of advanced intellectual practice. In this climate, whiteness, in engineering and other disciplines, functioned as an unmarked category.

Fortunately, the democratic impulses awakened in some Americans by World War II were not weak. The postwar period was a moment when identity politics began to coalesce for many historically marginalized groups in the United States, laying the groundwork for powerful assertions of African American pride and solidarity in the 1960s. No racial category would remain entirely unmarked for long. Furthermore, the violent and nonviolent protests in the 1960s helped spread the idea that black collectivity need not equate with group subordination to white interests, as proponents of a new educated "leadership" for the black community had hoped might occur in the 1940s and early 1950s. Heightened appreciation of black identity and heritage empowered minority citizens to bring changes in many public institutions, including education at all levels.

The extent to which the civil rights movement transformed the racial profile of science and engineering, or other realms of knowledge production centered in the university, requires much more study than it has thus far received. In the 1950s, sociologists of education outlined the role that education played in maintaining race and class divisions, especially in Britain. Their work identified the raw political power of credentialing systems in connection with admissions, curricular, and other educational standards. As the United States faced increasingly disruptive episodes of racial unrest after 1960, especially in the cities, local, state, and federal governments tried to answer those discontents with programs ostensibly geared toward black economic empowerment. In those programs, such standards played an ever greater role in maintaining the whiteness of science and engineering. Associations of rigor and selectivity in U.S. university engineering programs held fast through these activist decades to emerge unscathed in the 1980s when the nation returned to a quieter, more conservative, atmosphere. The conservative functions of merit in engineering and the contributions of such standards to black underrepresentation in that field through the 1960s and 1970s are the focus of Chapter 4.

CHAPTER FOUR

Opportunity in the City

Engineering Education in Chicago, 1960–1980

LEADERS of the University of Maryland system, clinging in many ways to the state's segregationist past, resisted what they considered to be a tidal wave of integrationist legal activity in the late 1950s and early 1960s. But it would be an error to assume that conservative anxieties in Maryland foretold a sea change in black opportunity in the nation. Stasis on many social equity matters characterized even the country's northernmost states. In urban settings, such as Chicago and Detroit, race conflict reached a fever pitch in the sixties, as did calls for educational reform. Poverty in American cities seemed to many to be a problem of racial inequity. Yet even as Northern cities embraced technical and commercial modernization, bent as was Maryland and much of the nation on participating in a modernized, globally important postwar economy, these communities too systematically deprived black Americans of the chance to participate in such development. As both legislative efforts and violent protests against racial discrimination gathered steam after 1960, many old structures of educational and occupational exclusion nonetheless remained intact. In this and the following chapter, I turn to Chicago during the 1960s and 1970s to trace significant yet circumscribed efforts at minority inclusion in engineering. These efforts ultimately reflected a limited commitment to diversification.

The goal in shifting focus to an urban case is to build on our understanding of engineering education as both a source and product of political stasis, a body of knowledge and practice rooted to a degree in a racially stratified worldview generally resistant to radical social change,[1] and to continue the task of explaining institutional rigidity in the face of shifting economic conditions and social upheaval. Historian Robert Fogelson, analyzing the origins and impacts of race riots on U.S. cities in the 1960s, wrote that bureaucratization, professionalization, and centralization in urban police, schools, welfare, and housing operations perpetuated discriminatory conditions for many black urbanites, even as individual prejudice may have been diminishing.[2] Higher education, and specifically engineering education, as institutional projects may be interpreted through this same historical lens. By the end of the 1950s, Chicago faced an intersection of economic and social challenges to which university growth, especially in technical areas, seemed one solution. First, within just a few years of the end of World War II, the city began to see the encroaching rust-belt deterioration of traditional Great Lakes heavy industries such as steel, shipping, and manufacturing. Leaders in business and politics shifted public and private resources toward aerospace, computing, and other emerging technology fields. New land use policies and tax structures supported the creation and retention of higher-tech industries in Chicago. Higher education could provide an impetus for industrial development through the provision of research and skilled personnel on which such modernization depends. Chicago's political and educational leaders had close ties to Washington in an era when federal money flowed relatively freely toward higher education, and the city garnered significant resources for its colleges and universities. In the decades following World War II, state and local legislators, civic leaders, and educators designated two inner-city institutions of higher learning as particularly important centers for the advancement of Chicago industry: the private Illinois Institute of Technology (IIT) to the south of the Loop (Chicago's central business district), and the public University of Illinois at Chicago (UIC) just west of the Loop. In the eyes of their proponents, both schools would offer new or expanded engineering degree programs and research facilities to help lift the city above its economic malaise, and both institutions became through the 1960s and 1970s loci of regional commercial development.[3]

At the same time, Chicago's leaders saw inner-city development as a hedge against social disruption, a growing concern for many majority urbanites. The city had seen a steady expansion of the poor, largely black neighborhoods near downtown since the 1920s; by the early 1960s, growing black militancy expressed by black Chicagoans worried many corporate leaders and politicians. Distressed by the centrifugal flow of more affluent city taxpayers and consumers toward Chicago's suburbs, and terrified by race riots and political protests throughout the 1960s, this tightly knit group marshaled legal and economic resources for projects that retained enterprises near the ailing Loop. Private interests attracted city, state, and federal support for renewal schemes intended to help rescue manufacturers, hospitals, universities, and myriad smaller downtown businesses. My focus here on engineering education and research shows how ideologies of technological and commercial modernization supported both one another and the race-based economic inequities instantiated by so much urban renewal activity.[4] IIT and UIC each at one time considered, but rejected, suburban locations in favor of urban sites, encouraged in that choice by city officials and businessmen seeking to protect their downtown investments from urban decay. However, while they may have secured local land values, the development of these two schools made limited contributions to individual economic opportunity in science and technology sectors for Chicago's booming college-age minority population.

The growth of IIT and UIC reinforced some long-standing racial inequities in the inner city. A mayoral commission, reporting on the 1968 race riots on Chicago's west side, saw many injustices in ghetto life for the city's black youth, but accepted as a foregone conclusion that most of that community's minority youngsters "of course, will never go to college."[5] IIT and UIC fought that diminished expectation to certain a degree, but nevertheless contended with counter pressures for institutional survival similar to those that shaped some educators' work in Maryland. It is crucial that we recognize innovative programs for minority engineering students at both schools but that we also acknowledge the systemic obstacles faced by their creators.[6] Certainly IIT offered increasingly prestigious degrees to students from around the country as it grew in the 1960s, and many Illinois families found affordable education for their children at the new UIC campus. IIT ultimately crafted novel and highly influential minority engineering programs in the early 1970s, but these made conventional measures of student selectivity a central feature and thus projected only limited reforms to occupational exclusion. Neither IIT, with its heritage of working-class education, nor the engineering programs of UIC, part of a public land-grant university, radically restructured opportunities in booming technical fields for the economically disadvantaged black students who made up a growing portion of Chicago's population. This was especially true for those African Americans living in the central city.[7] Despite the presence of energetic and progressively minded administrators and educators at both institutions, to the extent that the two schools helped to revitalize Chicago's industrial economy and prospered as institutions, they did so without maximizing the prospects of most minority citizens in technical occupations.

The histories of IIT and UIC in this era represent a powerful mutual influence of technological and social planning in modern industrial polities, and most importantly, expose prevailing ideas about what needed fixing in ailing American cities and what did not. Chicago's leaders in the 1960s and 1970s responded to rampant minority poverty and social unrest with a set of interventions that preserved inequitable political and economic institutions, and universities fit this moderately reformist template.[8] Many educators appealed to the necessity for "high standards" in technical education, associating expanded opportunity with diminished rigor. As social scientists of the time pointed out, this kind of functionalist ideology detected certain economic and technological needs in society that education of a particular kind could fulfill, and foreclosed recognition of educational (or employment) standards as reflective of social hierarchies.[9] That association of rigor and selectivity also laid the groundwork for the very heated arguments about affirmative action of the 1980s and beyond (explored in later chapters). It firmly cemented the idea that successful professional engineering practice is determined by the pre-college experiences of the person undertaking that practice. The "pipeline" idea gained credence through the civil rights era and supported the theory that elementary and secondary, more than postsecondary, education required address if America wished a more equitable occupational profile. That enduring formulation is placed here in its historical context.

In important ways this intersection of race and economic agendas as it played out in Chicago in the 1960s and 1970s echoed that seen in Maryland during the 1950s, when powerful politicians and industrialists there sought to answer social demands for change while keeping existing (white) economic advantages intact. In both settings, civic leaders associated expanded engineering education with regional and national economic development. The University of Maryland and Chicago's universities could ride a wave of excitement among influential citizens about technical expansion that defined industrial modernization as an unalloyed public good. But by the 1960s, educators faced conditions that were in many ways more urgent than those earlier encountered by Maryland's educators. First, the association of race and economic inequities were now being articulated in both official rhetoric and public opinion with unprecedented vigor. Organized resistance to Chicago's discriminatory housing and education policies made it very difficult to define *any* educational enterprise as entirely unrelated to racial inequities, let alone publicly supported education.[10] Activism among Chicago's Hispanic and Native American communities further charged the atmosphere, drawing new attention to the association of ethnic identity and economic disadvantage in the city. The demographic demarcation of engineering from other subject areas that were diversifying more rapidly (such as education and some humanities fields) became a focus for many engineers in and beyond the academy.

That demarcation became particularly pressing as the 1960s wore on and increasing numbers of Americans began to question whether technology was best viewed as practice devoid of political responsibilities and consequences. Public anxiety about science and engineering that had centered on the atomic bomb grew with environmental and anti-war movements in the 1960s (the impact of that tension on engineering research will be the focus of Chapter Five, which explores the political nature of academic engineers' research choices). Dramatically new ideologies regarding science and society were heard in U.S. universities; however, engineering teaching and research in Chicago projected social changes that fell short of complete reform.

Illinois Institute of Technology: Change and Stasis

IIT came into being in 1940 because Chicago had, since the mid-nineteenth century, been one of the nation's leading centers for the manufacture and transportation of consumer-oriented goods. For nearly 100 years, food processing, toolmaking, and publishing enterprises grew steadily in the city alongside heavier industries such as steel, chemical, and lumber production. In the 1890s, religious leaders and philanthropists founded Armour Institute of Technology and Lewis Institute for the training of Chicago's South Side youth, offering unprecedented economic opportunities to tens of thousands of working-class and immigrant families over the following decades. The two "street-car colleges" gradually shifted from vocational subjects to more formal, four-year technical degree programs, with Armour in particular aspiring to the status of the Massachusetts or California Institutes of Technology.[11] Officially linked in 1940, Armour and Lewis looked toward a greatly expanded role in the region's economy, reasserting their identification with a neighborhood far less affluent and commercially developed than the Loop but very much at the center of civic concern. Now titled the Illinois Institute of Technology, the school conjoined ambitions to promote itself as both a center of technical training and research and a major player in Chicago's urban renewal. For IIT, staying on the South Side and helping to redeem that distressed area seemed a way to contend with challenging local economic conditions and to participate in promising national technological trends for the good of school and city.

IIT's new Board of Trustees—a group representing influential business and civic segments of the city—immediately sought to build a large new physical plant, beginning with fifty acres. The school wished to maintain its comprehensive night school program and did not want to move beyond easy commuting distance for people working in the Loop or in nearby industrial districts; elevated trains, busses, street cars, and main traffic arteries already served the Armour site. With other large South Side institutions such as Michael Reese Hospital, IIT believed that in rejecting relocation to the suburbs of Chicago, it could help regenerate an area in severe economic decline, finding support for its own growth in the city's eagerness to address urban decay.[12]

That eagerness had been long building. Chicago had experienced a huge influx of African American residents with the great waves of Northern migration after 1910, and many of the newcomers settled just south and west of downtown. In a steady physical deterioration that began after major race riots in 1919, the near South Side witnessed the departure of white property owners from the area and their conversion of buildings to rental properties in which the landlords invested little time or money. Restrictive covenants prohibited blacks from renting in many other parts of the city and by the 1930s, the city government's systematic neglect of South Side schools and infrastructure had made the neighborhood one of the most dilapidated slums in the nation.[13] IIT, under the leadership of its first president, Henry Heald, wished to play a major role in the reversal of this dramatic decline. Trained as an engineer at the University of Illinois in Champaign, Heald had begun his career teaching at Armour Institute. He rose to become dean of students there, and by the age of 34, president, guiding the merger with Lewis to form IIT.[14]

Predicated in many ways on Heald's and the Board of Trustee's close ties to Chicago's government, IIT's involvement with urban renewal efforts was facilitated by an immense "bureaucracy of blight" that Chicago's civic leaders established in the postwar years. Heald rapidly assumed a leadership role in the Metropolitan Housing and Planning Council's program to "stand and fight" center-city decay through the use of public money as a foundation for private development. An ever increasing array of new federal programs intended to address severe urban housing shortages provided a source of such money, and in 1946, IIT joined with Michael Reese Hospital in inaugurating the South Side Planning Board to enact renewal plans for its immediate neighborhood.[15] Like many other new urban planning boards around the nation, the South Side board urgently recommended extensive changes. As historian Daniel Bluestone has characterized such renewal projects of the era, the more dramatic the visual alterations to "blighted" urban neighborhoods, the clearer the ostensible evidence of reform. Under these plans, old slums around IIT would be torn down, their residents relocated (ostensibly to purpose-built, public housing projects, although such construction was frequently delayed or resulted in substandard accommodations), and the cleared land designated for use by commercial or service concerns.[16] In 1950, a promotional brochure for IIT's new campus announced that, "In the heart of the United States' largest slum area, Illinois Institute of Technology is building its new modern utilitarian campus." This brochure also stated that both the Lewis and Armour institutes had originally stood nearby in what were once "desirable residential districts," and that to stay in the area, IIT would need to rehabilitate a now "run-down" and "degenerated" setting. To do so would bring about a "general uplifting of the entire south side."[17]

With an IIT administrator often at its head over the next decade, the South Side Planning Board helped the Chicago Housing Authority channel federal and private money into a series of residential redevelopment projects. In the zero-sum game that is inner-city land use, in which land is limited (and, in most planners' view, each tract can only be used for a single purpose), the growth of hospitals, universities, and commercial concerns depended on the elimination of vast residential tracts considered by planners to be blighted.[18] By 1954, IIT had designated 110 acres of housing in the neighborhood for demolition, planning their replacement with a modern complex of classrooms, laboratories, student and faculty housing, and utility structures. Through the 1950s some twenty-five buildings were erected. IIT's agenda for civic and institutional improvement in this era continued through the efforts of Heald's successor as president, John Rettaliata, who came to IIT in 1945 from research positions at Allis-Chambers, a major producer of machinery in Milwaukee, and the National Aeronautics and Space Council, predecessor to NASA. Rettaliata joined IIT as a professor of mechanical engineering, eventually becoming director of that department, vice president for academic affairs, and then dean of engineering. In 1953, he became IIT's second president. Rettaliata carried forward the vision of IIT as a source of technical personnel and economic strength for the city, promoting the research centers, primarily Armour Research Foundation and Institute of Gas Technology, developed there in the previous decade.

Defining Blight

New teaching and research buildings for the burgeoning enterprise were designed by Ludwig Mies van der Rohe (director of architectural programs at IIT for many years) in a starkly modernist style, complemented by functionalist dormitories and apartment buildings designed by the firm of Skidmore, Owings and Merrill.[19] The resulting campus remains today a strikingly austere and elegant assemblage of landmark International Style buildings and it takes a major effort of imagination to summon a sense of the busy neighborhood of shops, apartments, and saloons that once stood here. The economic deprivation common in that lost neighborhood is indisputable, but it is important to understand that the means by which IIT's new campus—known as the "Technology Center"—eradicated that poverty represents a particular solution to the problem of urban decline, with particular social costs.

The bold functionalist profile of IIT's new buildings equated cultural achievement with technical prowess. The buildings overtly displayed the technologies of their construction, with undisguised structural elements and unfinished industrial materials replacing a traditional campus aesthetic of brickwork or decorative stone carving. The contrast in color, shape, and surface with Armour's original masonry building next door (which still stands) is notable. Through this aesthetic choice, IIT's new buildings gave the impression of new technical knowledge applied in place of old-fashioned skills. As an editorial in the *Chicago Tribune* declared in 1965, the new campus acted as "a physical demonstration of the school's capacity to achieve."[20] Yet, for all its aesthetic dynamism, in some ways the campus functioned as traditionally as that of any Ivy League school, declaring the institution to be removed from its setting in practical operations and cultural import. In the early 1960s a pamphlet for new students entitled "Living at IIT" celebrated the school's "Metropolitan Campus," stressing IIT's proximity to the shopping and museums and other cultural amenities of downtown Chicago (a "mere 15 minutes away"), but also the on-campus banking, gas station, and recreational facilities that clearly signaled the separation of IIT's population from that of the older surrounding neighborhood. This kind of self-containment may have been a conventional choice for universities in this period, functioning *in loco parentis*, but IIT in so defining its independence also maintained its isolation from the surrounding urban life.[21]

By creating an attractive and prosperous enclave near downtown, IIT assured itself of support from Chicago's major business leaders and politicians, all seeking to halt economic erosion and channel federal funds into inner-city renewal projects. But IIT also needed to compete in the higher education market and by the 1960s it faced severe pressure from two other schools in the Chicago area offering, or soon to offer, engineering degrees. The private Northwestern University in Evanston, ten miles north of the Loop, had an established reputation in engineering research and teaching. The University of Illinois' new campus, while still in the planning stage, loomed large as a future threat to IIT's traditional student base and share of regional research funding. As tuition costs rose through the decade, more students turned to Illinois' public colleges and universities, and private donors followed that shift, further drawing income away from private institutions. What is more, by 1964, defense sector layoffs made engineering less attractive as a career choice for college enrollees than it had been for some time.[22] These trends left schools such as IIT faced with the dilemma of either raising tuition further or depending on federal aid that was by no means guaranteed.

President Rettaliata argued at the end of the 1960s that IIT's continued existence depended on the school establishing itself as a first-rate teaching and research institution, not in the local training and service mold of earlier decades. But in light of the nation's post-World War II technological growth, and Cold War inspired emphasis on cutting-edge research and development, this was an increasingly expensive proposition. Rettaliata justified this costly ramping up of IIT's standing with an apparent reference to the huge new UIC campus, by now in operation just a few miles away: "[O]f the many arguments based on the pluralistic traditions of this nation, let me briefly present just one. . . . The first class private schools serve as benchmarks against which the tax-supported institutions can gauge the quality of their product."[23] That elevation of IIT's standing repositioned the school for Chicagoans who once might have turned to it for affordable, accessible technical study. It stood now as a more elite institution, reasserting social divisions of the larger culture through its equation of selectivity and rigor.

The direction chosen for IIT at this juncture represents a complex agenda in higher education of the civil rights era. In 1965, the school undertook a massive campaign to raise $25 million from private sources for the establishment of some twenty new degree programs and new laboratories, classrooms, and residential buildings to accommodate the predicted growth in enrollment. The academic budget increased 70 percent, and the number of fulltime faculty grew by 41 percent in the following five years. Research was no less of a focus. In the course of the decade, IIT pumped millions of dollars into research facilities intended to fulfill the needs of Midwestern industries in chemical, mechanical, and other technical fields. Ten million dollars were spent on the IIT Research Institute (IITRI) Research and Administration building alone. The campus expanded and further exerted IIT's claim on the neighborhood as a new entity divorced in style and function from the district's earlier incarnations. As Bluestone summarized, "planners were determined to carve out a single-purpose academic campus where engineering students would be insulated from the very society they were being educated to serve."[24]

And yet, the school celebrated its economic contributions to the local community in which it stood, understandably trying to justify its use of public resources and revealing in the process some of the priorities of those controlling that money. In a 1966 report, IIT administrators saw the institute's economic support to the South Side as including wages, the purchase of materials and services, construction costs, research funding, and student spending. The economic impact that the report's authors called "perhaps IIT's greatest" was the entrance of IIT graduates into business and industry; with "70 percent of them beginning careers in the Midwest, 50 percent of them in Chicago." However, few IIT students came from the immediate neighborhood and fewer still graduated to nearby employment.[25] Overall, IIT's increasingly selective criteria limited admissions of students from weakly performing South Side high schools. The replacement of poorer, often black residential neighborhoods with large, white-run service institutions such as universities crafted an association of poverty and race that helped to solidify discrimination even where blame for poverty was not explicitly placed on its victims. In an informal address around the time of his inauguration, Rettaliata said, "Located as we are in a *highly industrialized* area, our programs reflect the people and industry in the community" [emphasis added].[26] This representation of the South Side as industrial in character, and by extension an area having few residential tracts worth maintaining, is consistent throughout Rettaliata's tenure, if less dismissive of the area's original residents than the claim of a 1950 IIT brochure. That publication stated that the school took over leadership in local slum rejuvenation because "absentee ownership precluded active leadership from within the blighted area."[27] While it is possible that renters represent overall a more transient population than property owners, IIT's leaders were participating in a larger program of defining black residents as somehow ineligible for (or at best, unlikely to achieve) the rights and privileges of community membership, even as these leaders imagined themselves to be addressing the worst effects of urban poverty.

After riots rocked the city's poorer districts in the mid-1960s, Chicago leaders moved to a somewhat more validating understanding of ghetto discontent than they had previously displayed. Mayor Daley's Chicago Riot Study Committee, like similar commissions in other troubled cities, disavowed violent protest but saw good reason for the "dissatisfaction" expressed by poor, black Chicagoans.[28] Civic participation was imagined to be a remedy for both individual dissatisfaction and community deficiencies; the Riot Study Committee saw "a growing feeling of black pride and social consciousness" among young Chicagoans as a potentially constructive force in poorer neighborhoods, if channeled properly: "Progress will only be made in the solution of ghetto problems, if there is effective communication between ghetto residents ... and expanded opportunities for indigenous leadership to participate in shaping decision and policies which affect their community."[29] This varies considerably from the tendency of riot analysts a few years earlier to blame urban riots on outside agitators, an explanation that effectively denied both the distress and potential civic contributions of rioting residents.[30] The Chicago Riot Study Committee nevertheless concluded that black youth had so little pride in their homes or communities that these young people felt they "had little to lose by destroying it all."[31] However much other portions of the report may have recognized the role of systemic discrimination in the discontent of Chicago's black citizens, this finding portrayed the rioter not just as marginalized, but as misguided and self-destructive. All these official judgments about poor Chicagoans' character and civic potential created a context for the limited progressivism of minority education efforts in the city. That IIT understood local demographics is without question: students and faculty helped create and run community service buildings and playgrounds in neighborhood housing projects and made significant contributions through those efforts. Yet, whether through direct and conscious judgments made by IIT's leaders about poor city residents, or a desire among those leaders to please a city government and business elite on whom they depended for financial resources, IIT defined its neighborhood's original residents as part of the problem of blight and not of its solution. Those citizens were largely consigned to a world apart from the technical and economic expansion that IIT represented.[32]

University of Illinois at Chicago: Building Circle

If IIT's leaders worked hard to claim selected public functions for their private institution, leaders at UIC defended an identifiably privatistic vision for Chicago's new public university. Problems of racial conflict and expanding poverty, and a commercial sector that felt itself to be at risk from both, led to the creation of a state university branch whose social agenda was more like its private counterparts than might be expected. IIT justified its receipt of public urban renewal funds by emphasizing its role in fending off urban blight; UIC sought to convince Chicago's business interests of its value to established patterns of economic advantage. It did so by perpetuating differential opportunities for Chicagoans of different class and racial backgrounds. UIC differed from IIT in some important respects, but ultimately the public school, too, faced systemic pressures that led it to preserve relatively conservative social functions for engineering education in the city.

Establishing its new campus about four miles to the north and west of IIT, the University of Illinois began in 1965 to offer through its urban branch a public university venue for the training of engineers in Chicago. The university had been chartered as a land-grant school in 1867 in the rural town of Urbana, 140 miles south of Chicago. Its roots were agricultural, but by 1910 the school had a stellar reputation in engineering and conducted research for a wide range of commercial and civil applications. During World War II the university established a small commuter branch on Chicago's Navy Pier, primarily for returning veterans. This eccentric structure (routinely flooded when large ships passed by) offered only two-year programs and was never intended by the university to be a permanent campus. Plans for a permanent branch solidified in the 1950s, however, when demographers indicated that the number of Chicagoans likely to seek public higher education was about to boom. In the expanding postwar economy, working-class white families, recent immigrants, and black families who had arrived in the decades-long northern migration sought improved economic conditions for their children. The civil rights movement encouraged the desire among minority citizens for public higher education. This was not a population that could easily afford to board children at the downstate campus. To many in Illinois a greatly expanded upstate facility for commuter students was an exciting prospect.[33]

The choice of site for the new campus just west of the Loop involved tremendous political infighting. As historian George Rosen has recounted, many Illinois legislators supported the idea of locating the campus in a distant suburb of Chicago where land was cheap, even if it was beyond the reach of public transportation. Mayor Richard J. Daley, however, advocated a site near downtown Chicago previously earmarked for future federal housing renewal projects. As part of the Near West Side Conservation Project, started in 1955, that renewal effort already had support from the federal government. With close ties to the Kennedy administration in Washington at the start of the 1960s, Daley could realistically expect to draw even greater federal funding for a new public university campus on the site intended to serve 20,000 students. Daley believed that such a project would work to halt encroaching economic deterioration of the Loop area, as white residents and businesses around the city center moved to the suburbs and poorer black and Hispanic citizens took their place. Just as was the case at IIT, boosters promoted the so-called Circle campus as a hedge against further white flight because it would replace houses and apartments with a large, permanently funded institution. Unlike IIT's South Side setting, this neighborhood held families largely of Greek and Italian heritage, with most of its residents living well above the poverty line in housing that few considered degraded. But planners nonetheless considered this "buffer zone" around the Loop to be at near-term risk of decay. Legislators approved a 118-acre site, known as the Harrison-Halstead parcel, for the new campus, incurring the dismay and sometimes fury of the neighborhood's soon-to-be-displaced residents but fulfilling the desire of Daley and his supporters for a protective corridor around the central business district.[34]

The new campus was physically ambitious. Like those involved in designing IIT, UIC's planners meant to impose an entirely new civic and aesthetic identity on the site. The first phase of building, between 1963 and 1965, supplied fifteen buildings for some 9,000 students, including a twenty-eight story administrative building, a science and engineering building, library, classrooms, and a lecture center. The second and third phases, continuing through 1969, created more facilities for the sciences and the humanities and allowed for enrollment of up to 20,000 students on the site. The university's trustees called upon the architects of Skidmore, Owings and Merrill for this academic project, as well, and the firm designed a complex of low-rise concrete buildings interspersed with taller office and classroom towers. The style was modernist, but here executed with a less clean or delicate feeling than that of Mies's IIT vision of the previous decade. The most visible material was concrete, with most surfaces left rough and unpainted in the powerful new "Brutalist" style of the day. Thick columns and beams predominated at and above eye-level. Dean of Engineering George Bugliarello, arriving at UIC in 1969, described the scene: "[A] modern campus, and one that has the courage of its own opinions, no attempts at imitations, Gothic, or what have you. A bold statement."[35]

As at IIT, planners chose visual conventions for this campus that they hoped would convey a particular mindset relative to traditional university architecture and to residential structures previously standing on the site. Designed as a commuter school, UIC provided no dormitories but still configured itself as a setting separated by aesthetic feeling and layout from nearby shops and restaurants. By 1965, a 100-acre campus of futuristic structures and elevated walkways—a setting labeled by one architectural critic as "hard, unyielding, and vast in scale"—had eradicated all signs of the lower- and working-class neighborhoods from the site.[36]

Much as IIT had described part of its purpose as local economic uplift, so UIC's supporters issued assurances that local minority-owned businesses would be granted contracts for university work. The school imagined a community presence beyond the provision of higher education. This was, after all, a public university, and from its inception UIC worked to establish extension education and community welfare programs for Chicago's inner-city residents.[37] But the broader impacts that public spending on the university might have on Chicago presented a far more contentious matter. The essential conservatism of Chicago's leadership during the civil rights era is well documented. Early in his tenure Mayor Daley famously abandoned his connections to Chicago's black Democratic voters in order to serve the white electorate more directly. In so doing, he reinforced economic and geographic divisions of the city's population along racial lines.[38] As historians Alan Anderson and George Pickering have noted, a large sector of business leaders and politicians in Chicago shared a "civic credo" that denied the seriousness of racial inequities and supported the maintenance of existing institutions such as those behind Chicago's educational, employment, and housing provisions. At the same time, many influential Chicagoans perceived the conduct of electoral politics, city agencies, civic boards, and local universities to be rooted in the tenets of liberal democracy and therefore, despite ongoing signs of unrest, to be guarantors of some eventual solution to racial inequalities. Drawing on theories of American race relations that found hope for black assimilation in the experiences of European immigrants earlier in the twentieth century, even some progressive thinkers could accept with some ease the limited efforts of Chicago's leaders to correct race-based poverty.[39]

If Chicago itself was split along lines of race, Illinois in its entirety had long lived with a political split between its northernmost, urban constituencies and the rest of the state, home to less ethnically and economically diverse communities. Local patronage systems in Cook County collided with agendas emerging from the state legislature in Springfield. Unlike IIT, the state university branch in Chicago had to contend in very direct ways with downstate Illinois leadership, a cohort even less invested in urban racial and economic reforms than the city's elite. Many state leaders from outside of Cook County supported increasing resources for the University of Illinois' Champaign-Urbana campus, and a smaller branch at Springfield, rather than new expenditures for a Chicago campus, mindful of their own constituents' educational interests. To those uninterested in advocating for inner-city development, suburban Chicago locations for an upstate campus of the University of Illinois offered more value for money, and further detracted from the appeal of a downtown campus.[40]

To meet the demands of both upstate and downstate legislative audiences, proponents of an urban University of Illinois campus articulated a range of justifications for its presence in downtown Chicago. A college of engineering in the new location could promise an influx of customers for local businesses and improved land values, as would any division of the new campus. However, as a site of technical teaching and research it could serve local industry and attract local corporate resources to the inner city, thus contributing to the city's economic growth in obvious ways. It could also offer an affordable means of professional training to city residents as the only public, degree-granting engineering program in Chicago. And finally, if any of these benefits were to accrue to downtown Chicago the city would become less of a drain on state resources. Eventually, the Illinois legislature did support the creation of the Chicago branch with an engineering division; however, the direct provision of engineering skills to economically disadvantaged Chicagoans—enacting the goal of compensatory education in technical fields—did not survive as a lasting part of UIC's agenda. As universities took on the role of interveners in worrisome patterns of urban decline, offering cities a means of renewal through their well-capitalized production of new knowledge and trained workers, the question inevitably arose of who would be eligible for that training. The desirability of an "urban mission" for the Chicago campus was debated within the university, the city, and between Chicago and the rest of the state. It was eventually rejected as a directive for some divisions of UIC, including engineering, by both administrators and faculty.

Whom to Serve: Constituencies at IIT and UIC

Through the century's middle decades, as it redefined its role in Chicago from a gradually advancing vocational school to a premier "Technology Center," IIT saw its mission shifting from the preparation of skilled technicians of many types to the training of professional engineers and scientists. In keeping with that mission, IIT steadily diminished the role of its Evening School. For many years before and after World War II, IIT served the majority of its students through evening classes, making technical education available to working people. In moving a much larger proportion of its resources to daytime teaching, the school hoped to serve both a more selective student population and one drawn from a much wider geographic area. By 1967, 33 percent of freshmen entering IIT came from outside Illinois, a far cry from the older model of a commuting school targeted at working-class Chicagoans.[41] Preparing self-study materials for the American Society for Engineering Education (ASEE) in 1964, IIT engineering faculty noted as one "essential" goal that a "careful distinction be developed between what is and what is not the profession of engineering . . . the distinction between an engineer and a technician." They stressed also the increasing role of engineering graduates in management and the public sector, thus echoing a devaluation of technological skills and a commensurate promotion of scientifically and mathematically informed practical research that had been a staple of established academic engineering departments since World War II.[42] With this new agenda came increasingly selective admissions policies. Further, a shift of emphasis to graduate study meant that by 1973, 37 percent of IIT's degrees were being granted at the graduate level.[43] The creation of nondegree curricula for local industry employees promised attractive sources of revenue. These involved classes offered by closed-circuit television at company headquarters and special topics classes held on campus outside of working hours. All represented a further commitment of resources to a market remote from IIT's immediate neighborhood on the South Side.

At the same time, IIT's leaders were experiencing clear pressures to increase minority participation in engineering. Calls for greater minority inclusion in all kinds of occupations across the American economy were supported by multiple forces. Since the Eisenhower administration, Washington had been studying systematically the problem of racial exclusion from white-collar professions. Under President John F. Kennedy, equal employment opportunity became a programmatic emphasis. Some 270 companies pledged to find, hire, and promote minority employees through the government's 1962 "Plan for Progress," a program representing more than eight million jobs. Among these were some of the highest level science and engineering positions in the country, overseen by employers who claimed to be frustrated by the lack of "qualified minority group individuals—Negroes in particular."[44] With Lyndon Johnson's Great Society programs, a vast federal commitment to education promised unprecedented representation of minority Americans in lucrative occupations.[45] Anxiety about social unrest was unquestionably a growing reason behind such initiatives. The violence of inner-city racial protests through the second half of the sixties prompted President Johnson to form the National Advisory Commission on Civil Disorder. The commission was headed, at least nominally, by Illinois Governor Otto Kerner, and reported a dangerous racial divide arising from economic inequalities in American cities.[46]

Urban renewal and related federal programs directed at higher education built on and reflected impulses within academia to address this problem. In 1964, the Ivy League colleges joined together to send one, joint representative on a "talent hunting trip" to "Southern Negro High Schools," a gesture noted in a new publication of the American Council on Education (ACE) entitled *Expanding Opportunities: The Negro and Higher Education.* Technical disciplines made similar efforts. In the early 1960s a number of well-established engineering programs began pursuing the "most needy" students, including programs at the University of Wisconsin and Cornell. The U.S. military instigated "career guidance for disadvantaged students" in metropolitan areas in collaboration with urban engineering programs in the early 1970s. The private sector also undertook the induction of minorities into technical fields, with the Carnegie Foundation supporting partnerships between black and white colleges with engineering programs beginning in 1960.[47] By the early 1970s, the National Academy of Engineering (NAE), like similar bodies in medicine and law, began systematically to study issues of inclusion along lines of race and gender, while educators' and philosophers' demands for a "humanization of technology" likely also encouraged awareness of social justice issues in engineering and engineering education.

In 1973, the task of diversifying American engineering along ethnic and gender lines gained a new foothold when the chairman of the board of General Electric (GE), Reginald Jones, examined minority hiring patterns at the company and discovered satisfactory growth in numbers of minorities hired, but a generally low level of employment and mobility among those hirees. GE soon helped to found a conference board of some 350 American corporations with the objective of increasing the number of black engineers from "400 to 4,000 by the middle of the 1980s." At the same time, prompted by GE's activities, the NAE established a Committee on Minorities in Engineering, comprised of an advisory board of industry representatives and a paid group of professional personnel. The committee conducted studies and workshops, and offered advice to federal agencies, congressional staffs, the White House, and scores of universities and private corporations.[48] That decade also saw the founding of the National Advisory (now Action) Council for Minorities in Engineering (NACME; created in 1973) through efforts of NAE members, in which some of the largest firms in the nation participated. In consolidating industry efforts to strengthen minority inclusion in engineering, NACME brought both greatly increased funding and public visibility to business interventions in higher education. With significant support for minority engineering education forthcoming from larger philanthropies such as the Alfred P. Sloan Foundation, many businesses saw good reason to contribute to such related initiatives as the National Fund for Minority Engineering Students.[49] When IIT sought a new president following John Rettaliata's resignation in 1973, the school chose an educator with an active role in many of these developments.

Minority Engineering Programs at IIT

Thomas Lyle Martin, Jr., who became president of IIT in 1974, was an electrical engineer who had previously been a dean of engineering at Southern Methodist University and other institutions. He rapidly corrected what he saw as IIT's "collision course with [financial] disaster" by bringing up enrollments,[50] but it is clear that the school's Board of Trustees also saw in Martin a man with sensitivity to contemporary social issues then troubling inner-city institutions. Martin had recently served on the steering committee of an NAE symposium on increasing minority participation in engineering and he was a tireless supporter of the Committee on Minorities in Engineering and NACME, chairing the former later in the decade. Martin's immediate willingness to be associated with diversification at IIT is evident in a single illustration: the opening, full-page photo of "Highlights of the Year at Illinois Institute of Technology" in Martin's first "President's Report" of 1974, is of black teenagers testing a model bridge in IIT's new six-week summer Early Identification Program (EIP) for Chicago high school juniors.[51] The choice of this photograph over the traditional campus vista or gleaming portrait of some new high-tech instrument positioned the new administration as one willing to advertise its involvement with social change.

Minority engineering programs initiated at IIT under Martin reveal both reform aspirations and limits to such projects during the civil rights era. Sociologists and education policy makers had, since the late 1950s, steadily stepped up their inquiry into how best to correct differential school performance among black Americans, who, it was widely perceived, tested lower than majority students and ended up in less lucrative and rewarding occupations. There was no clear sense by the 1970s that one factor was determining that inequity; analysts offered biological difference, family and cultural influences, economic standing, and school characteristics as competing explanations. As the civil rights movement gained momentum, analysts increasingly looked away from individual cognitive or behavioral shortcomings among poor-performing minority students toward organizational or social-structural explanations.[52] But as affirmative action gained credence as a possible solution to race-based inequities in higher education, many academic disciplines, including engineering, were placed in a difficult position. In arguments that were far less strident but still reminiscent of those made in earlier decades, academics now asserted the importance of maintaining current standards of college and university education in the nation. Martin conceptualized minority underrepresentation in engineering as a three-part problem involving address at the pre-college, college, and career level. He maintained a nuanced and ambitious approach to the problem, but in each area, he articulated solutions that preserved the essential content of engineering education and practice while aiding individual students with a carefully defined array of services.

Upon his installation, Martin rapidly built up initiatives recently begun at IIT by Director of Placement and Co-operative Education William R. Smith and Dean of Engineering and Physical Sciences Peter Chiarulli. These initiatives had set up both a pre-college co-op experience for interested IIT applicants and small programs for high schoolers interested in obtaining some familiarity with math and engineering. The U.S. Department of Health, Education and Welfare had given IIT a grant in 1972 which led to the hiring of Nathaniel Thomas as Minority Co-op Co-coordinator, and upon arriving at the school Martin found a receptive audience for his priorities. New summer programs and school-year weekend programs gave students from some of Chicago's poorest high schools a visceral sense of technical practice, often through the close involvement of regional industries. Martin saw student motivation as an "exceedingly complex factor" that required an integrated address of self-esteem and experiential issues. By the mid-1970s, IIT was offering eight-week summer programs for minority pre-senior high school students, a senior year math and physics component, summer employment for pre-college freshmen entering IIT, and various support services for minority freshmen. All of these activities had both academic and social components. In building these EIPs, Martin meant not only to identify minority students likely to succeed in the field, but to establish among those students a sense of identification *with* engineering as an occupation; "human relationships" were central to his pre-engineering vision.[53] Another obstacle to minority inclusion in engineering, however, was one that Martin saw as less readily subject to correction in individual students through any effort IIT might make. This was the lack of mathematical skills among many students graduating from inner-city high schools.

While Martin stressed the importance of high school training in mathematics and related skills, implying that improved high schools would improve minority representation in engineering, he also rejected the idea of IIT providing college-level remedial work that might supply those skills. In 1976, speaking on "maximizing black potential in science and engineering," Martin said: "While remedial programs are vital to the overall minority education picture, we must work to destroy the image that remedial programs and minority education are synonymous. And, we must establish the difference quickly if we are going to make any headway in achieving a more realistic allocation of funding and effort to minority engineering education."[54]

Martin saw the underrepresentation of minorities in engineering as "a clear waste of talent and human resources" and recognized the personal disappointments and challenges faced by students who encountered discrimination. But, partly through that emphasis on "highly personal" and "individual" interventions for minority students, IIT's programs for minority students nonetheless reinforced a conception that limited universities' responsibility for social change. Despite its origins in progressive social ideology about race relations, that concept helped to define poverty as a matter of individual disadvantage, not of structural inequities.[55]

Martin openly disagreed with analysts who stressed standardized test scores as the only meaningful gauge of student potential. Although things had changed greatly since a 1964 IIT study reported that Chicago's secondary schools were of such low caliber that "only placement exams can be trusted" in the assessment of applicants,[56] Martin wished to leave crucial elements of engineering curricula and professional practice undisturbed. When IIT began in the early 1970s actively to recruit black students, the school emphasized that only qualified, rather than qualifiable, students would be considered, exempting the institute from the work of bringing disadvantaged students up to speed academically. The EIPs at IIT drew upon the top 10 percent of interested high school juniors, and in 1980 the school claimed that the success of its Minorities in Engineering program resulted from the admission of only "a select special type of student." Surely Martin's subsequent point that, "once these kids learn someone cares and that they can do the work, they breeze through," justified in a certain sense his warning to other institutions that "lowering entrance standards, developing a less demanding curriculum, or providing remedial work after college entrance are *not* solutions to the problem of getting minority students into engineering schools" (original emphasis). His outlook matches that of most other programs deemed at that time to have been successful, and many universities by the mid-1970s had ended programs aimed at enrolling "high-risk or under-prepared blacks."[57]

On the broadest pragmatic level, the rejection of remedial functions for higher education had a reputational function for university departments. Prominent high-tech firms (including Honeywell, employer of some 5,000 engineers and scientists) could make the case around this time that many minority engineering graduates did "not come up to the standard provided by the major accredited engineering schools." This argument had internal logic, as employers added that those graduates were "definitely not up to the level required for them to have any chance for success in the work we have to do."[58] Martin's proscription protected existing conceptions of what college-level work in engineering should look like. Its exclusion of lower-level, remedial classes for incoming students prevented what might be construed as a loss of instructional rigor at IIT. This approach required minority students to be fully prepared *prior* to arriving at university and to be chosen for admission in light of that preparation. For proponents of this view, college engineering curricula were black-boxed as socially pristine and devoid of political import.

As sociologist of education Michael Young summarized in 1970, by focusing on the "characteristics of failure," educational analysts could build support for programs that might improve the participation of disadvantaged populations. But most progressive voices in education in this period could not detect what Young calls the "socially constructed character of education that the students failed at." Student failure remained an indication of individual deviance, not of systemic ills.[59] This emphasis helped to naturalize the distinction between accomplished and failing students, further justifying any rewards for stronger scholarly performance.[60] It also meant that programs for minority students might be deemed successful if they proved effective for only small numbers of students. Martin's approach to the problem of student retention follows this pattern. Minorities generally had a far lower retention rate in the discipline than majority students: in the late 1970s, 77.8 percent of white students versus 54.8 percent of minorities finished engineering degree programs.[61] Faced with the retention challenge at the college level, Martin advocated a continuation of the kinds of efforts seen in pre-college interventions for minority students: co-op jobs that brought both added income and real-life experience to students, advising about educational options, and mentoring to overcome social difficulties. Academic intervention, however, was not an option. That approach deflected inquiry about some basic features of engineering curricula, including about the rarity of remedial or protracted coursework. Thus, while Martin's interventions worked very well for those students who were deemed eligible for participation (in 1976, only 5 percent of students who entered IIT under the EIP programs subsequently quit their degree training), they also reiterated selection criteria that continued wider patterns of racial underrepresentation in engineering.[62]

Because immediate successes were achieved with these highly selective approaches to minority inclusion, it requires some historical distance to understand why they did not lead to larger numbers of minority engineering graduates over time. Under Martin's administration, IIT quickly became a leader in the enrollment of minorities in engineering fields: 10 minorities were counted among 80 new freshmen in 1973; 40 were counted among 174 freshmen in 1974; and 40 to 50 minorities entered each year through the rest of the decade.[63] IIT publicized these programs widely and gained the support of businesses, some of which donated substantial funds to the projects and held IIT up to other schools as a model. The NSF funded many of these initiatives at IIT, praising them as exceptionally successful.⁶⁴ On the national level, NACME gave large grants to universities in this period to aid in minority outreach, recruitment, and the provision of supplementary scholarships. The United States saw a rise in minority freshmen enrolling in engineering, from 2,249 in 1973 to 11,116 in 1981, but thereafter, the growth ceased, and then numbers of minority freshmen in engineering declined through the 1980s. Today, with many MEPs still based on the admission of "qualified" rather than "qualifiable" students, the representation of African Americans in university engineering departments remains low for reasons we may trace to this inaugural period and the type of strategy represented by IIT's programs.⁶⁵

As a set of answers to occupational inequity, interventions such as IIT's could only grow to a certain scale before reaching a plateau. NACME researchers, trying to explain the cause of the stagnation in the early 1980s, found that: "the pool of minority students graduating from high school with the credentials to enter engineering school had not changed during the late 1970s, thus 1980–81 marked a saturation point in the recruitment effort."⁶⁶ It is exactly the nature of that "pool" that requires historical analysis if we are to understand the ongoing underrepresentation of African Americans in engineering. If a student arrived at the end of high school without having achieved visible and conventional measures of academic success, he or she was not invited into the world of higher engineering education. This was disproportionately the case with minority students who made up the majority at underfunded, inner-city high schools. With an emphasis on outreach and recruitment, IIT's programs brought high school students into the places where engineers worked—factories, laboratories, and classrooms—and gave them a taste of the work engineers did. In so doing, the school successfully identified talented students who might not otherwise have been made aware of engineering as a career, but these programs defined talent somewhat narrowly as a certain level of competency in math and science. What is more, some educational specialists concerned with race equity had found that disadvantaged social background could inhibit a student's pre-college achievement. Thus, to mandate the admission of only qualified and not qualifiable students put the burden of integration on a part of the education system that patently was not doing its job.⁶⁷

At the same time, analysts recognized that some students involved in race activism displayed characteristics conducive to personal achievement: "hard and persistent effort" and group goals that were "both difficult and realistic." These minority students, committed to collective action and achievement through cultural or political initiative, might have lower grades than some other students, but upon closer examination might be seen to have intellectual and character strengths.[68] That sort of alternative evidence of occupational promise was not visible to the majority of educators seeking future engineers, dependent as those educators were on conventional metrics of student achievement. Since at least 1970, some educators have suggested that pedagogical stress on learning processes, rather than outcomes, may best equip students for work in the sciences. Neither "content accountability," which requires that instructors cover certain content (usually assessed through standardized tests), nor time pressures encourage such open-ended inquiry in the classroom, which some observers associate expressly with increased retention among women and minorities in the sciences.[69] Finally, we may understand that the American stress on the measurement of students' innate ability is itself culturally specific. In China and Japan, for example, students, teachers, and parents approach difficulty with mathematics instead "as a problem of time and effort," an outlook that would likely encourage far more attention to institutionalized learning conditions than does a focus on an individual's capacities.[70]

Disincentives to Diversity

Martin championed the idea of an ethnically integrated academic culture, and IIT achieved that integration for a certain period under certain institutional conditions. Yet IIT's minority initiatives fit a template of individual achievement that in the long run could not correct an overall pattern of black exclusion from engineering. What would have happened if institutes of technology and universities, given enough support and sanctioning by employers of their engineering graduates, took over the task of remedial education? Schools could conceivably maintain sound material standards for engineering but train students over longer periods, with greater provisions for remedial instruction. Even if such instruction costs more than existing approaches, why have educators and policy makers historically not found it worth expending such resources to correct a social inequity? The answer lies in part in the disciplinary organizations from which IIT's College of Engineering and other engineering programs derived their credibility in the 1970s and which function similarly today.

Reputations are at once abstract cultural qualities and the products of concrete social and institutional structures. Most immediately, engineering programs have to answer to academic accreditation bodies that seek certain course distributions and student performance levels, and funding sources that seek certain rates and types of research output. Internally, engineering faculty answer to standards for tenure and promotion that reflect disciplinary conventions which elevate innovative research and devalue teaching and service. No individual instructor, department, or university can afford to dismiss the reward systems that maintain these values.[71] In the pedagogical realm, entities such as the Accreditation Board for Engineering Technology (ABET) have insured that most engineering degree programs have developed curricula of fairly similar lengths, and with nearly uniform amounts of mathematics, basic sciences, and humanities. While the accreditation system does not dictate exact course content, it does direct the emphases and general level of instruction. These features of engineering curricula translate into encouragement for certain student constituencies. Curricula of longer duration, for example, that might accommodate remedial instruction, would depart from such standards.[72]

As it aimed toward a higher status among engineering schools, IIT saw that engineering programs around the country were becoming increasingly rigorous, further discouraging remedial or extended curricula that might aid disadvantaged students seeking engineering degrees. A 1979 report issued by UNESCO summarized the previous decade as having brought a general shift upward in the level of course content, until engineering education in the United States had become, "in a sense . . . an upper-level two year programme." The report described the greater depth of mathematics and physics now required in engineering as occupying students' first two years of college, and an increase in graduate instruction forcing "the implementation of a large number of specialized senior-level courses as prerequisites" to graduate study.[73] In 1973, Paul H. Robbins, executive director of the National Society of Professional Engineers, identified a shift in professional engineering whereby proficiency in measurement, surveying, drafting, and preparing blueprints had been displaced by an "entirely different capability" centered on higher-level technical and research skills. Those more basic tasks were now being performed on the level of "engineering technology"—a nonprofessional level of training usually provided by two-year institutions. IIT pursued the most rarified niche for its graduates at the upper end of the engineering profession: research and development positions rather than administrative or sales jobs, or those production jobs now being labeled as engineering technology tasks. This entailed curricular choices weighted toward more specialized and sophisticated content.[74]

Although such decisions limited some interventions for disadvantaged students, one should hesitate to associate such pragmatism with any particular political ideology. IIT leadership showed no inclination toward the idea of limiting governmental interventions in private education, for example. In the late 1970s, Martin could point to consistent support for his efforts from deep-pocketed patrons such as Standard Oil, Illinois Bell, and Inland Steel, but he saw that the efforts of private institutions and their corporate supporters to correct discrimination in engineering could not succeed without much greater federal involvement than was currently the case. Despite outlays of individual and local energy and funds that Martin characterized as "huge," minority engineering programs were "headed for turmoil" if more federal support was not found. Where are "the NSF, Office of Education, Department of Labor and Office of Science and Technology Policy?" Martin asked.[75] Nevertheless, study of Martin's and similar approaches to racial diversification in engineering unveils a robust set of arguments against compensatory education. In 1978, Martin approvingly cited comments made by Lindon Saline of General Electric on the "minimum demands" that minorities must make of their professional employers. These included the provision of "real work" ("no tokenism") and an assurance of "no free rides," each demand intended to ensure that any offer of employment would be an authentic one through which the hiree found satisfying and contributory work. In this way, tendencies among employers to lower the bar for black applicants would be countered, and the quality of new, minority entrants into technical fields might be controlled, ostensibly for the good of employers and employees alike. Martin's own assessment of the Committee on Minority Engineering noted that, "We are achieving affirmative action to right past injustices, but without reverse discrimination," invocations of merit again serving to foreclose discussion of the processes by which merit is defined.[76] Concurrent developments nearby at the University of Illinois' new College of Engineering will show this same powerful linkage between ideals of rigor and selectivity in this period.

University of Illinois at Chicago: The Educational Assistance Program

Once its planners defined it as a comprehensive university to be located in Chicago's inner city, UIC faced paired tasks: first, fulfilling an "urban mission," as many called such broadly inclusive educational agendas, and second, achieving credibility (if not prominence) across a wide range of academic disciplines. University of Illinois system President David Henry designated Norman Parker, head of the Mechanical Drawing Department at Urbana, as the primary planner for the system's new Chicago campus. As the new facilities opened, UIC initiated in its thirteen-story Science and

Engineering building degree programs in Energy, Information, Materials, and Systems Engineering. This was an unorthodox organization of subjects by technical function rather than by established discipline. Engineering students chose from among twenty-two "areas of concentration", and all graduated with a Bachelor of Science in Engineering—a degree that intentionally was not to be specified further. This curricular structure, departing from the format usually associated with engineering accreditation guidelines, promised students "more flexibility in choosing a career" than did traditional programs. Employment analysts of the period reported that subfields within engineering could experience heightened demand at different times, and UIC's planners may have felt that their graduates' ability to move between or meld specialties served both students and their prospective employers well.[77] Furthermore, students' concentrations might involve work in more than one department, encouraging interdisciplinary study. That loosening of boundaries among subject areas held the potential to support study and research on nontraditional topics, a curricular choice with political implications (discussed more thoroughly in Chapter 5). The initiation of an unusual curriculum design proved inimical to the long-term interests of the college, but at UIC's opening, the somewhat daring curriculum complemented an inclusive approach to engineering that made pedagogy and student preparation for industrial jobs priorities of equal or greater importance than faculty research.

At its founding, the College of Engineering operated within the generally very open admissions guidelines of the campus as a whole. UIC admitted all students it believed to have a "50/50 chance of success." An additional 10 percent of each incoming class was admitted without regard to high school rank or test scores through the University of Illinois' system-wide Educational Assistance Program (EAP), begun in 1968.[78] UIC's engineering division supported this effort to recruit and sustain minority students with lower college entrance scores, welcoming EAP students with apparent enthusiasm. In 1970 the College of Engineering hoped to have 150 of its anticipated 650 arriving freshmen come from the ranks of EAP students. The University of Illinois provided all of its EAP students with financial aid and counseling, as well as strategically reduced degree requirements and special classes in reading and study skills, rhetoric, and mathematics. UIC's College of Engineering provided its own EAP students with additional tutors and the use of a research laboratory as a study and social center.[79] Notably, while clearly concerned with minority inclusion in technical occupations, the University of Illinois established in Chicago only high-level engineering degree programs (albeit under unusual rubrics), and significantly did not choose to set up an Engineering Technology degree program. If successful, UIC's approach would bring a previously excluded cohort into competition for the most lucrative stratum of technical employment.

Support for this inclusive agenda continued under the leadership of Dean of Engineering George Bugliarello, who assumed that position in 1969. The College of Engineering had not had a permanent dean since 1966, shortly after the campus opened, and Bugliarello brought to UIC a dual focus on engineering pedagogy and the application of engineering to social problems. His advocacy of UIC's unorthodox organization of engineering subject areas signaled a disregard for some other rather significant professional conventions; throughout his tenure at UIC Bugliarello advocated courses for engineers in the history of technology and other new interdisciplinary subject areas. He co-organized an ambitious symposium on the history and philosophy of technology with the dean of the College of Liberal Arts and Sciences, Dean B. Doner, and published widely on those subjects. In 1971, Bugliarello chaired an early conference on "Women in Engineering." Two years later, he coauthored a sweeping NSF-funded study of interactions among "Technology, the University and the Community" in Chicago. That study infused questions of industrial development in the city with concern for class and race inequities. Bugliarello and his coauthors distinguished their study from earlier investigations that looked only at economic costs and benefits to a community without assessing the "long-range influence of a source of skilled manpower, or the socio-economic impacts on the graduates themselves."[80]

However, even before Bugliarello left UIC in 1973, forces conspired to reduce inclusivity in engineering programs there. First, changes occurred in the operation of support programs within the College of Engineering. In 1971 the college created an "Engineering Self-Help Center" that initially aimed to help minority students by providing orientation and tutoring services, but shortly evolved into a resource for "all engineering students with handicapped backgrounds or having difficulty with the curriculum." Any student helped by the center was expected to "repay" the service by in turn working as a tutor, advisor, or recruiter for the center. The shift to partial self-support for minority services may indicate the school's move away from its fullest commitment to minority recruitment. In 1973, Bugliarello reported an increase in minority students in the engineering college "from 50 to 200" since its opening eight years earlier. Yet, by 1972, the engineering college had itself declared that "the minority engineering student at the University of Illinois has been very unsuccessful in the matriculation process." The College of Engineering proposed to continue the admission of some students with "inadequate academic backgrounds," but also saw a need to "redefine teaching methods": "Through the Educational Assistance Program we have found that remedial courses are not what is [*sic*] needed to attack the problems of the urban education system in America."[81]

By the early 1970s, academic leaders at UIC had begun to detect an incompatibility between social interventions and the long-term success of the College of Engineering and of certain nontechnical divisions of the new university in which research had been made a low priority. It is necessary to look to the larger institutional culture to understand why the College of Engineering thought it necessary to reduce, rather than expand, its remedial interventions for minority students who were not reaching success through existing EAP operations.

A movement generally to increase the quality of scholarship and research on the UIC campus at this time gained momentum under Dean of Faculties Glenn Terrell, who saw UCLA as the ideal model for UIC. California's comprehensive public university had in large measure achieved parity with private teaching and research institutions in that state.[82] UIC's College of Engineering felt this disconnect between its founding and emerging identities intensely. Paul Chung, who became the dean of engineering in 1979 and remained in that capacity for the next fifteen years, looked back at the decade preceding his appointment and saw it as a period of "turmoil" in which "the campus had all but abandoned meaningful admissions standards" and "many of the competent faculty were leaving for better positions elsewhere." Others agreed that UIC's generous admissions policies had eroded its reputation among industry and government agencies and productive senior researchers.[83] If UIC as a whole was "adrift" in this era, as one journalist put it, the engineering school provided minimal anchoring. As of 1972, only eight of fifteen areas of concentration in the engineering college had been accredited by the Engineers' Council for Professional Development (later, ABET).[84] In many respects the university's social welfare commitment to the community, evident in dozens of outreach projects linked in varying degrees to its academic programs, did not diminish in the least. But the College of Engineering increasingly saw itself as having to function apart from such commitments, and opportunities for Chicago's disadvantaged citizens in technical education decreased measurably with the college's changing priorities.

The founding choice at UIC to operate with very open admissions policies had positioned it at what its leaders soon realized was a disadvantage among engineering schools. UIC's overall orientation toward citizen uplift meant that many departments chose to make research a lower priority than inclusive undergraduate instruction. This emphasis on teaching was initially an exciting goal for some early engineering faculty at UIC. As David Levinson, acting dean of the College of Engineering between 1966 and 1968 (just after the campus opened), explained, UIC's focus ran counter to the "present rage in university faculties . . . to concentrate on research and graduate education." But he projected an integration of "science and engineering" undergraduate training and a very flexible technical curriculum that would nonetheless make UIC engineering graduates particularly desirable to industry. An overemphasis on science, he felt, had recently diminished the real utility of engineering education: "metallurgical engineers are studying such things as color centers in irradiated alkali halide single crystals while the steel industry is becoming progressively more alarmed by the shortage of metallurgical engineers it faces."[85] A rejection of such highly specialized work in the College of Engineering, it was felt, would allow students to contribute not just to industry but to the improvement of modern life in general, addressing emergent human needs. Levinson also noted that the embrace of such unglamorous pedagogical agendas would "require skill and courage" on the part of faculty.[86] He understood that his vision represented a serious departure from customary institutional and career goals in engineering.

There were practical institutional consequences to UIC's paired goals of open engineering admissions and problem-centered learning. Within a few years of the college's inauguration, as Chung recalled, engineering faculty were steadily departing. Between 1965 and 1972, minority students without EAP support completed an average of 27.10 quarter hours of work per year while those with EAP support completed 32.6, but full-time university students were expected to complete 45 quarter hours of work per year. The apparent low caliber of students and lack of support for research together eroded instructors' morale and probably helped render some faculty ineligible for competitive funding. Urbana, where fiscal control of the university's Chicago branch ultimately resided, was bent on preserving resources and possibly prestige for its long established engineering research efforts; it gave little support to those technical research initiatives that did take root at the Chicago campus. One of the reasons that EAP engineering students at UIC had use of an entire laboratory for study space and socializing was that no faculty or funds were available to put it to its intended research uses.[87]

The incompatibility of social reform and institutional survival is the question, not the answer, if we want to explain the perpetuation of racial inequities in academic engineering. One EAP staff member at UIC recalls that EAP students admitted to the College of Engineering with low entrance examination scores often did not do well academically, and points out that some have made the argument that it was simply "unfair" to admit these students.⁸⁸ But why did the College of Engineering respond to this problem by adjusting its admissions standards upwards, rather than by changing the nature of its instruction to accommodate and assist low-scoring students? Why pursue different students and not develop new curricula to correct the incompatibility?

Answers to those questions for UIC's educators, as in any setting, derived from perceptions of acceptable risks and rewards associated with either course of action. As IIT's experiences indicated, throughout the 1960s engineering educators felt pressures from the professional sphere to achieve an escalation, rather than an attenuation, of undergraduate teaching standards, with graduate degrees looming as the new criteria for professional employment.⁸⁹ UIC's faculty watched these developments and contended with even greater pressures because of their dependence on the state's limited public funding for education. It may be impossible to distinguish where pedagogical standards ended and fiscal constraints began for engineering faculty at UIC, but without question their options were shaped in the larger political arena in which the campus functioned. In that sphere, monetary resources and social planning were mutually constitutive.⁹⁰

However genuinely reformist educators at IIT and UIC may have been through the 1960s and 1970s, they answered to boards of trustees rooted in local civic and commercial networks that forecast economic development predicated on social hierarchies of class and race. In an immense 1967 study of conditions and trends in "population, economy and land" commissioned by the City of Chicago and overseen by local business leaders, consultants predicted that a growing percentage of residents in coming years would be young, "non-white" males. While the report calls for "fair employment practices, job training and increased job opportunities" in both its general message and detailed recommendations, a different outlook for the growing minority labor pool is projected. According to the report's authors, these citizens would optimally be trained for and find employment in what was predicted to be a growing technological sector of the local economy, but at a particular level of skill and remuneration. The authors made explicit the level at which young minority men would contribute to Chicago industry: as "competent repairmen and mechanics to service new [aerospace and data-processing] equipment."⁹¹ The planners' report thus naturalized an association of technological expansion with a race-based hierarchy of skill and labor. The Chicago Department of Development and Planning foresaw as optimal a two-tier system in which white students and a few high-achieving black students would graduate into engineering positions, while the larger portion of the black labor pool would fill lower-level positions as technicians.

These assignments were solidified by the historic distinction between professional engineers and technicians already noted. Movement by an individual upward from supporting technician role to full-fledged engineer in American industry had been almost unheard of since the early twentieth century.[92] Some of Chicago's civic leaders in business, government, and education recapitulated intentionally or otherwise Daley's general resistance to significant redistributions of economic opportunity in the city. In the early 1960s, UIC's Norman Parker and IIT's John Rettaliata joined the mayor on the new Chicago Committee on Urban Opportunity (CCUO). In anticipation of federal programs soon to begin under the 1964 Economic Opportunity Act, the CCUO oversaw job training and experience and vocational guidance in such areas as primary education and social services. Resulting jobs were meant to include such positions as school aides or social agency staff. The idea was to find secure employment for "Youths 16–21 who come from communities which are economically depressed and culturally deprived. These youths do not possess the necessary educational prerequisites to participate in more advanced training programs."[93] Daley's insistence on local control of federally mandated anti-poverty programs generally kept the CCUO in the hands of known Democratic-machine operatives, a cohort openly opposed to political activism by those outside the machine. While some engineers, instructors, and administrators at UIC and IIT imagined a newly open world of occupational opportunities in a renewed Chicago economy, a radical diversification of professional engineering through public higher education was an unlikely possibility in such a milieu.[94]

Conclusions

Paul Paslay, dean of UIC's College of Engineering from 1976 to 1978, accepted the college's inclusive admissions policies ("Our campus has a particular problem, for it wishes to provide maximum opportunity in an urban setting to students who have received public pre-college training.") But with pragmatic concerns, he made it clear that certain standards were inviolable: disadvantaged students were to receive help in making up academic deficiencies, but only for "a reasonable amount of time [about a year]." And, faculty must be "prepared to fail unusually large numbers of students compared to engineering colleges with stricter admissions policies." According to Paslay, employers of UIC engineering graduates, "have a right to expect a level of technical competence competitive with major engineering schools. When we fail to provide this level of competence, we hurt the graduates, the college, and the profession."[95] The vulnerability that Paslay ascribes to the engineering profession (not to mention to engineering institutions) is telling. In actuality, admitting students in need of remedial support has no direct connection to the graduation of underqualified practitioners, only to the necessity of expending more resources on the education of those particular students if one wishes to maintain existing standards of professional qualification.

This selective view of how engineering might change to accommodate a more diverse workforce is echoed across the literature on the technical labor market. Discussing the small number of women in engineering in 1970, one study concluded that few women experienced "unjustifiable discrimination" when attempting to enter the profession. Instead, the report concluded, employers excluded women from engineering on the basis of "economically justified" reasons. These included the fact that women commonly married, left a region when a husband received a job transfer, or experienced "breakdowns in childcare arrangements." The study noted that "frequent job changing and interruptions are not conducive to developing expert command of complex and rapidly growing technologies."[96] The report explicitly says that women have an aptitude for engineering; however, the author does not question social patterns (women's customary responsibility for childcare, for example) that render women ineligible for professional status in engineering. The continued growth of the technology sector actually appears to *depend* on the preservation of inequitable social structures, again associating social change with technological regress and economic risk.

Crucially, the history of minority underrepresentation in engineering involves not only the plans of educators, politicians, and boards of trustees, but those of students and their families. In the EAP's first year of operation, only 6 percent of UIC's EAP students chose to enroll in the College of Engineering. The vast majority enrolled in the College of Liberal Arts and Sciences. The underrepresentation of blacks in engineering at IIT and UIC indicates not just institutional actions but perceptions and choices of minority students planning their education, as concerned educators at both schools and around the country recognized. Those choices made by individuals and their families, of course, reflect their perception of institutional and occupational conditions.[97] Majority students also have a political self-identity. Dean Bugliarello recalled with evident pride a campus-wide student strike at UIC, sponsored by the radical Students for a Democratic Society, during which engineering students defended "with bicycle chains" their right to attend classes.[98] Study of both minority and majority occupational aspirations can tell us much about deeply seated expectations regarding the social origins and roles of American engineers.[99]

But all of these self-concepts play out against an occupational structure that in the 1960s and 1970s saw many countervailing trends to civil rights and race activism.

By the end of the 1960s, the larger national outlook for engineering employment had also deteriorated, lending new motivation to those amenable to limiting minority inclusion in the field. With contractions in the defense industry and fears of a general recession on the rise, many engineers saw an imminent "no-growth" situation; it was no longer the case that "anybody who wanted to work could work." One member of the national Engineering Manpower Commission said that, in order to address this oversupply, "immigration, which supplies engineers, can be slowed down if needed."[100] Under such conditions, pursuing a measurably expanded domestic labor pool through racial diversification would not have made much sense without a much stronger impulse toward racial equity for its own sake.

IIT and UIC, as sites of technical work, are both like and unlike other city institutions of their era. As would any other private or public enterprise, the schools made priorities of existing economic advantages, with social reform taking a second place to survival. But unlike commercial enterprises and public service entities, engineering schools arose from a larger cultural milieu that customarily places technical work in a realm divorced from social factors. Definitions of rigor, objectivity, and material achievement placed on universities by accreditation and funding groups narrowed IIT's and UIC's options. The consequences of that narrowness for racial diversity are not coincidental. Christopher Newfield has described the development of American notions of meritocracy in the late nineteenth century in a way that helps explain the mixed progressive and retrogressive character of Chicago's engineering programs in the civil rights era. For Newfield, meritocracies function on the basis of uniform measurement of all citizens on the same scale, creating "an abstract monoculture" in which cultural, national, linguistic, "and every other type of difference" are relegated to "the status of subjective features irrelevant to rating a person's quality."[101] In some ways, academic engineering is the perfect "abstract monoculture," systematically elevating a single set of criteria for student eligibility over varied kinds of evidence of talent or drive. It is the sincerity of engineering educators' diversity efforts in the 1970s, through which they sought out and supported talented minority individuals who might otherwise have fallen through the cracks, that shows the regrettable limits of meritocracy as a liberal social force. In that era, when so many social reforms were being tested, the optimism of educators at IIT and UIC may have been understandable. It was not, however, enough to bring widespread change to the educational system.

The claims of fairness for meritocratic occupational systems have only grown louder in ensuing decades. One feature of the debate about affirmative action has been the insistence by its opponents on the removal of race from all considerations of intellectual merit for hiring or college admissions purposes; the postulate, as economist and educator Glenn Loury has summarized it, that "racial identity should add nothing to an assessment of individual worth."[102] This goal envisions a realm of practice simply stripped of social consideration, a handy ideal of civil and intellectual practice. Engineering, like science and math, customarily bases much of its utility on its freedom from "extraneous" factors, so it is easily enlisted in efforts to equate important intellectual work with the erasure of race. That function for technical knowledge and training in the late twentieth century becomes even more evident when the focus shifts from engineering pedagogy to engineering research in the urban university.

CHAPTER FIVE

Urban Engineering and the Conservative Impulse

Research at the University of Illinois at Chicago and the Illinois Institute of Technology

THE SMALL SCALE of minority engineering programs in American universities throughout the 1970s reflects the incompatibility of these programs with institutional survival. In order to thrive as sites of engineering instruction and research and to meet the expectations of funders and employers, universities put limits on their diversification ambitions. Their design of engineering programs for minorities, which generally precluded admission of lower performing students and the provision of remedial coursework, complied with long-standing ideals of institutional performance rather than with growing national sentiment for increased minority presence in white-dominated occupations. That sentiment, while measurably greater in this era than at any previous time in American history, went only so far as a force for institutional change. Locked into competition for students and support, universities and their faculties, including many instructors at UIC and IIT, believed that they could not afford to evade convention by accommodating a long-excluded constituency. To deviate from convention was to risk the loss of influence and legitimacy, and so African American students remained on the margins of academic engineering.[1]

How strong would civil unrest, race activism, and demands for equitable education have to have been to have shaken the standards established for engineering instruction and brought about the racial integration of engineering professions? In some ways, through the 1960s and 1970s American engineering departments displayed an unusual level of attention to the larger social conditions in which they functioned. Unlike the 1950s, when relatively few practitioners within technical disciplines voiced politically or socially progressive opinions as part of their professional work, this period saw the clear expression of liberal social agendas in engineering. By the late 1960s, technical educators and practitioners joined a growing number of Americans concerned with the social impacts of technology. Unlike technocratic impulses of the early twentieth century, some engineers now voiced grave doubts about the potential of technology to solve human problems. This was a trend that had its origins in the nature-venerating transcendentalist movement of the nineteenth century. It found momentum after Hiroshima and increasing force as environmental worries and anti-Vietnam War feelings brought to many college campuses an atmosphere of skepticism regarding technological expansion. Many scientists and engineers in this period promoted work on projects that fell outside customary industrial or military applications.[2] Urban universities in particular began tentatively to claim a role in the solution of social problems such as substandard city housing and health care, persistent underemployment, and pollution that envisioned an enlistment of technical faculties for those unconventional purposes. In courses for engineering students, and in research projects chosen by instructors (which informed undergraduate curricula and graduate research work), engineering educators introduced subject matter of a liberal bent.

But this conversion of engineering work from familiar industrial, civic, or defense applications to the production of socially informed structures and systems did not gain a lasting foothold. Programs at preeminent schools such as MIT, funded by the Ford Foundation, and at selected land-grant schools such as the University of Wisconsin did inaugurate the address of so-called urban problems. Other schools followed suit and a handful of these initiatives led to permanent programs or research centers. But most were gone by the mid-1980s, having disappointed at least a few of their sponsors. Most were not meant to transform engineering curricula or research, but rather to confine ancillary subject matter in special centers for urban research.[3]

The institutional marginality of socially informed engineering instruction and research is central to the story of racial inequity in Chicago's academic engineering programs. On one level, expanding the set of problems on which engineers might reasonably spend time (for example, by maximizing attention paid to substandard housing, transportation, and medical facilities as technical problems) would have justified increasing the pool of available practitioners by any means possible, including through the recruitment of women and minority students into engineering. On another level, by admitting matters of social equity as a necessary component of its work, engineering might have taken on the task of correcting race or gender-based inequities within its own corridors. Finally, with greater academic emphasis on social problems, minority students might have felt a greater interest in becoming engineers. The idea that marginalized citizens are somehow naturally interested in the technical problems of their own communities is problematic, and later I will address how this idea nevertheless achieved currency among many educators.[4] But it does seem likely that if disadvantaged students perceive a discipline to be socially exclusive and thereby resistant to diversification, and what is more uninterested in its own social profile and therefore averse to change, they will be unlikely to seek careers in that discipline. In all these ways, the rejection of a concerted social focus embedded habits of racial exclusion in American engineering.[5]

The approach to social problems that was exemplified in academic engineering of the 1960s and 1970s reflects an ambivalence toward social change not unlike that expressed by the development of large technical institutions in decayed urban neighborhoods. In both instances, Chicagoans with political influence and cultural resources chose to address race-based poverty with sincere but limited reform efforts. As Chapter 4 indicated, city leaders saw urban decay as a problem solvable by the provision of opportunities primarily to white land and business owners. Chicago's universities developed programs aimed at minority engineering students; however, these programs were kept on a small scale and engineering occupations remained the province of advantaged students, primarily of majority background. Similarly, members of the civic elite and educators, presented with the possibility that engineering might address urban problems, did not reject political functions for engineering outright, but confined those involvements to projects of a certain scale. These efforts would not present prospective employers with unconventionally trained engineers, nor would they provide cities with dramatically new technical resources for the solution of urban problems. Instead, they would adhere to existing technological priorities in American culture while reinforcing existing structures of opportunity in engineering.

As historian Jennifer Light has shown, a number of American scientists and engineers carried their wartime expertise to the challenge of urban problems as defense spending shrank in the 1960s. For many, the physical and mathematical analysis of complex ballistic and combat conditions seemed to parallel the untangling of the economic and social chaos then afflicting American cities. Experts previously employed in military areas sought to transfer operations research and related techniques to troubled urban centers. But that initiative did not achieve widespread success. These "defense intellectuals" faced difficulties of translating issues of social welfare into technical terms. That difficulty reflects both the limits of systems analysis as a means of capturing lived social experience, as Light indicates, and the conservative moral economy of late twentieth-century engineering.[6] Both IIT and UIC hired engineering administrators who had shown a pronounced interest in social issues, including in the case of Thomas Martin, very direct concern with achieving racial diversification. UIC's innovative dean of engineering in the early 1970s, George Bugliarello, imagined active interchanges between Chicago's universities and its housing, public works, and transportation planners.[7] But, as with student admissions standards, there operated in this area a similar selection process, whereby both the private and public schools chose a certain level or type of engagement with social issues.

In this period a two-tier system of engineering teaching and research emerged. With a few exceptions, that system associated conventional subjects of interest to the military and industry with more prominent schools, and a focus on urban or other social issues with engineering programs of lesser status. For many Americans, that division resonated with the idea that minority students in engineering would naturally gravitate toward problems that disproportionately affected minority citizens. Identifying these and other race-based assumptions shows that ostensibly progressive engagements with the technological aspects of social problems, however sincere those engagements may have been, were not substantial enough to assure long-term reforms in engineering teaching and research.

It is possible to perceive the limitations placed on academic engineers' study of social problems by universities and their sponsors, and by engineers themselves, as a demarcation activity, in the mode of sociologist Thomas Gieryn's work on the boundaries of science. Gieryn shows how scientists' declarations of what is and what is not science (the rejection of "discrepant and competitive maps of the place of science in the intellectual landscape"[8]) have much to do with definitions of who is and who is not eligible to be a scientist. If practitioners consider an activity to be outside the sphere of science, then those who engage in that activity are not scientists. In this chapter, I will describe the process through the 1960s and 1970s by which engineering professions legitimized certain subject matters for research in response to an unstable social climate. The work of academic engineers in Chicago in particular shows cognitive habits as well as institutional choices that reflect political distaste for radical social reform.[9] The academic renunciation of urban engineering problems has followed the lines of a conceptual division: a firewall between technical and social thought or behavior. This firewall has taken the form of negative judgments about socially informed practice in the classroom or workplace. In terms of teaching, this old antagonism, described by a number of historians, ebbed and flowed from the inception of formal engineering training in the mid-nineteenth century in the United States.[10] For some years, philosophical conflicts between classical and scientific educators shaped competition for limited resources inside the modern university. Engineering programs throughout the first half of the twentieth century, eager to find managerial employment for their graduates, often included some social science training to address the "human factor."[11] However, by the mid-twentieth century, academic engineering had proved its utility to the military-industrial complex; management came to be characterized by more technical skill sets; and in technical education settings of the last four or five decades, humanities have often been treated as add-ons to engineering curricula at best, and expendable at worst. This trend, when examined at all by humanities educators, is commonly attributed to student schedules being "over-filled" with technical requirements. With a more focused causal analysis, the broader political forces behind this diminishment of liberal arts education in engineering may become clear.[12]

Urban Problems Defined

In 1979, a UNESCO summary of a five-year study on the global role of engineers depicted the history of American engineering as one of constant transformation, in which the discipline had continuously adapted itself to best serve national needs. Starting out in the service of an agrarian, then mechanical, and ultimately industrial national economy, by the mid-twentieth century engineering had adapted to practices that were "technologically advanced" and "heavily based on science," as was necessary for the application of engineering to aerospace, automation, and electronics functions. The UNESCO report's author, Lawrence Grayson, saw contemporary engineering instruction as a culmination of that timeline: "Today, curricula are beginning to include substantial amounts of the humanities and social sciences, as engineering is being applied to health, education, urban development, pollution control . . . and other socially related problems."[13]

The idea that engineers might tackle the problems of increasingly crowded cities and burgeoning industrial development was not new. In the early twentieth century, Progressive Era voices supported the role of urban engineers in everything from sewer design to the preparation of school lunches. As early as the 1890s engineering educators conscientiously prepared their students for the highest managerial and political positions in the country, with coursework in the classics, rhetoric, and ethics included alongside that required in science, math, design, materials, and more specialized technical subjects. The majority of engineering programs stressed to at least some degree their graduates' ability to communicate, to identify the greater good of a community, and generally to forward the largely unexamined causes of improved human civilization or the "benefit of mankind."[14] What Grayson was sensing as a new development in the late 1970s, however, was that many Americans had become increasingly sensitive to the idea that they were living in "changing times," in part because of technological developments that had brought about detrimental conditions such as urban crowding and pollution.[15] Unlike Progressive Era voices that may have blamed industry and commerce in some unspecified way for the discomforts of city life (as did advocates for city parks, for example), those of the later twentieth century produced explicit and elaborate criticism of technical spheres. Not all analysts of urban problems blamed economic and infrastructural decline in U.S. cities on technological expansion, but commentaries often included complaints about smog, outdated transportation systems, and decaying housing stocks. Fueled by anti-war sentiments that identified objectionable foreign policy with the conduct of military research and development, critics blamed technology as a body of knowledge for a range of social ills. Many observers, like Grayson, called with new urgency for solutions to those problems from a range of sectors that included the technological.

Meanwhile, for educators as well as the public, the idea that universities would play a generally increased role in the correction of urban problems gained currency through the 1970s. This shift arose as federal funding flowed into urban schools through such agencies as the Department of Housing and Urban Development, the Department of Labor, and the Office of Economic Opportunity (the same funds that had made the growth of IIT and UIC in this era possible). A trend toward decentralization, which created local branches of these agencies, was meant to make them more responsive to local conditions. In empowering local civic elites, this trend gave urban universities further encouragement to become involved with government interventions in city affairs. In the mid-1970s, the American Council on Education (ACE) held a series of regional conferences on the subject in which several themes emerged. First, urban affairs were of increasing importance to the economic and social well being of the nation as a whole. This sentiment was unquestionably a response to ongoing racial unrest in Los Angeles, Detroit, Chicago, and elsewhere, which had created an atmosphere of fear among white civic leaders. But with a constructive and potentially radical perspective, the ACE conferences further found that while American universities had not yet done so in any systematic way, with sufficient government and private support academia could potentially contribute to the solution of urban problems. And, finally, the ACE conferences found that this role was especially appropriate for public universities operating in the tradition of land-grant schools.[16]

All these proposals seem to support a democraticized application of national resources and a mustering of national intellect and energy toward unassailable goals. If any of these findings were to result in actual application by American universities, however, some complex and fairly controversial aspects of the project (none of which the ACE conference addressed explicitly) would have to be resolved. These include adjustments of student admissions standards (as discussed in Chapter 4) to bring new economic opportunities and skill sets to inner-city residents. Federal granting structures and academic tenure and promotion patterns would require revision, as well, to assure reward for untraditional, socially informed research and faculty service.[17] Underlying any of these changes are implications of truly radical reform that might have shaken even moderate conservatives, particularly those well beyond the academy. Because urban decay disproportionately affected nonwhite Americans, the application of engineering to urban affairs would bring engineers into play in ways that might challenge the economic bases of racial inequality. If engineers turned their efforts away from market-based applications of technology, they might instead devise affordable housing for the poor, expanded mass transit, and solutions to smog and the difficulties of solid waste disposal. This shift in emphasis not only could result in the delivery of improved services to disadvantaged urban residents, but it also imagined a considerable proportion of engineering work occurring apart from conventional measures of economic growth.

Such schemes touched on the larger social movements that questioned the products and values traditionally associated with engineering in the United States; for example, the Appropriate Technology and environmental movements. Historians have very recently begun to recognize concerns about technology within certain strands of the civil rights movement, as well.[18] In San Francisco, engineers working in the public sector expressed acceptance of this new interdisciplinary agenda that sought to bring social concerns and technical expertise to bear on one another. One engineer, who served as a general manager for the Bay Area Rapid Transit District (BART) in the early 1970s, described that mass transportation project as resolutely melding technical and "non-engineering related factors":

The engineer, devoting his professional skills and time to socially oriented projects such as he did with BART, does not anymore have the luxury of considering the prevalent political, environmental, and social factors to be entirely someone else's domain. He must, to be effective, maintain an open and communicative mind to ideas continually being provided by other political, social, and environmental protagonists.[19]

The author implies a tendency on the part of engineers to disdain integrated work, but sees no lack of ability or adaptability. He projects ultimately a significant contribution by engineers to the successful, socially sensitive completion of BART. In Chicago, where the prevailing atmosphere was not so clearly one of progressive change, those who designed and operated university engineering programs faced choices about their degree of involvement in such integrative movements. Cast in a favorable light by some academic engineers in Chicago's universities as "occupational social accountability," and by others as an inappropriate burden sure to bring the demise of engineering colleges or even of the entire profession, the social applications of engineering were subjects of constant debate at both IIT and UIC in the 1960s and 1970s.[20]

That debate resulted in a contained effort at socially oriented engineering, confined to the practices of a small number of instructors in specialized courses and in the operation of dedicated centers or institutes. IIT had grown through the 1940s and 1950s through federal and state largesse aimed at urban renewal, but the majority of work done within its laboratories and classrooms did not attempt to build on liberal social expectations. The work at this institution was geared toward existing economic structures in which industrial and military research agendas were predominant. For industry, IIT offered expertise pertinent to regional and national businesses. At UIC, a mission of public education and a legacy of land-grant service to broad swatches of Illinois' citizenry collided with a growing understanding of engineering as the basis of corporate expansion. In both settings, social issues that affected the inner city were not a major focus of technological address.

The Minimization of Social Research Problems in Engineering

If postwar institutional commitments were to be the basis of new research foci in the 1960s and 1970s, neither IIT nor UIC could easily have shifted their technical research to local concerns. Sporadically since the early 1950s, IIT's engineering faculty had undertaken research with potentially direct benefits to the neighborhood in which it stood. In one such case, in 1951, the civil engineering faculty, under contract with the Federal Housing and Home Finance Agency, investigated "cost saving by using better methods of construction and design of [public housing] apartment buildings." This direct address of Chicago's social problems was something of an exception, however. Although IIT did far more consulting to industry on engineering topics as a whole than did Northwestern University and even the University of Illinois at Urbana (which was geographically remote from Chicago), that work tended to support either local businesses or federal projects, not low- or nonprofit projects.[21] Not surprisingly, university involvement with local industry tended to reinforce existing business priorities. When area firms such as Motorola sought technical instruction for their employees, IIT offered a wide range of special courses on its own campus and at the company's suburban plants. Then, as today, it seems unlikely that customized courses for industry would introduce controversial or unexpected content.[22] In the 1960s, city business leaders made two predictions: a steady diminution of Chicago's standing in traditional heavy industries such as primary and fabricated metals, machinery, and stone, clay, and glass; and imminent technological change that would necessitate for business reliable access to the newest knowledge. IIT, a private school heavily dependent on industrial patronage, understood the practical necessity of serving those interests.[23]

As the U.S. Department of Defense built up spending for research and development during the Cold War, IIT also understood that some of its greatest research opportunities lay in the federal sphere. Since 1936 the Illinois Institute of Technology Research Institute (IITRI; first opened as the Armour Research Institute) had operated beside IIT as a separate, nonprofit center, much as did comparable institutes at MIT and Stanford University.[24] Such enterprises were established to encourage regional economic growth in technical areas that might be too expensive or too mired in confidentiality or proprietary issues to be easily undertaken inside the universities. After World War II, these research institutes increasingly took on tasks set by the Department of Defense and other federal agencies. IITRI maintained flexibility through the rest of the century, successfully moving between military and civilian spheres as the economy demanded. However, its directors never intended it to serve primarily local interests. Indeed, by the early 1970s a full 95 percent of IITRI's funded research was supported by federal agencies such as the National Science Foundation or National Institutes of Health. Nor did the research institutes at IIT, MIT, or Stanford necessarily make "major intellectual or teaching contributions" to the universities with which they were affiliated. In its early days IITRI relied on IIT faculty as part-time researchers, but that pattern faded by the late 1950s as completely separate staffs evolved in the research institute and school. The administrative separation of universities and their affiliated research institutes had political ramifications. With the isolation of the two kinds of operation, technical faculty who were less focused on profitable applications could not shape the research of the institutes. In any case, there is reason to believe that most IIT faculty were focused on national rather than local work. By the 1970s, IIT was repeatedly making the association of institutional "excellence" with national and international influence, claiming that "excellence transcends geographic boundaries."[25] If nothing else, a national audience meant a larger market for researchers' skills than a single city or region might offer.

Public universities, reliant on legislators' confidence in the schools' ability to build regional economies, had in some ways an even greater stake in claiming the broadest possible economic utility. Maximizing geographic reach for the sake of research revenue could easily be justified if one considered the university apart from its community. But contributions to be made by UIC could also be delineated by its urban location, distinguished by its founders from the Urbana campus on the basis of geography. These self-definitions vied with one another through the first decade of UIC's operation, reflecting the ambiguous agendas of many administrators and faculty members.

George Bugliarello, dean of engineering at UIC from 1969 to 1973, oversaw with energy engineering professor Harold Simon the creation of a report on the "regional role" of engineering colleges for the city of Chicago in 1973. Those commissioning the report, titled "The Impact of the University on its Environment," asked how engineering could best "satisfy the needs" of the city. The report's authors listed several possible answers. Conceivably, engineering programs could address the demands of industry by training personnel in fairly narrow technical specializations. A different emphasis would produce engineers trained in the economic and managerial aspects of city operation. A third outlook projected increased participation by engineers in the address of a "broad spectrum of social problems" faced by modern American cities. The report treats each possibility with equal seriousness, pointing out no incompatibility among the three priorities. The authors acknowledged difficulties with each, to which I will return. But the idea of engineers crossing disciplines and taking on tasks conventionally associated with the social sciences seemed a reasonable vision. The report was meant to encompass all engineering colleges in the city, not only UIC's, and envisioned Chicago's universities and corporations "evolving together in a mutually beneficial way," citing as an example the happy partnership of the chemical engineering department of the University of Delaware and its neighbor, the DuPont Company. But adherence to the land-grant ideal, wherein public rather than private universities answer the majority of research needs of state industries, can be seen in many of the report's projections. After all, the University of Illinois had operated the notably successful Engineering Experiment Station at Urbana since the turn of the century, and it must certainly have seemed to many in Illinois that similar work undertaken in downtown Chicago was guaranteed even greater success. The development of research geared to the public interest most broadly conceived, and not merely to those of large corporations, is central to the authors' vision here.[26]

To illustrate the potential contributions that engineering programs might make to Chicago, Bugliarello and his coauthors drew up an "Engineering College Impact Matrix" that laid out the range of services to be performed by engineering colleges and the constituencies for those services. In terms of services, this grid lent an equivalence to the "diffusion of skills and knowledge," "research," "public service," and "community activities" undertaken at the school. As for "groups affected" by academic engineering units, alongside cells for "students" and "faculty" and "industry," the authors included a slot for "shopkeepers," suggesting that urban universities really did reside in established communities to which engineering programs should pay attention. The report repeatedly acknowledged that the "general public" experiences the "impact" of engineering colleges, and cited Bugliarello's own notion of an imminent "Socio-Technical Phase" for engineering in which current aims of education broaden and the "laboratory becomes the region."[27] Clearly, however, this was an ideal, rather than a real arrangement. The report observed that the city's own departments and committees (such as those in charge of Chicago's aviation, building and zoning, transportation, and public works) which might have been expected to benefit from engineering research had "practically no professional contact with the engineering colleges" of the region. Cities of the day, the report explained, "have immense and urgent problems which could greatly benefit from advice and assistance from the universities—but which the universities have largely been unable to provide."

Unquestionably, the report's authors had UIC, then less than ten years old, in mind when they stressed that "emerging colleges [that are] still seeking a direction for a principal thrust" might not be hindered by "inertia," and thus could still respond to "local and national needs" in new ways. That inertia was, in some senses, the result of a trickle down of administrative priorities to the level of engineering laboratories and classrooms.

Through the early 1960s the availability of funding from the Department of Defense so outweighed that of any other source that academic programs and corporate research and development priorities simply skewed toward military projects.[28] However, when the defense sector's influence began to shrink, social problems still did not achieve a notably higher profile in the work of university engineering departments. Instead, political priorities affected choices of engineering research problems, guiding educators and administrators in private and public universities away from a focus on urban issues despite conspicuous interest in such issues within the scientific academy.

Faculty at IIT, a private school with aspirations to compete with the nation's highest ranking technical institutes, undoubtedly would have known about a growing movement to convert defense-centered research facilities to civilian areas in a handful of high profile research settings. Historian Matthew Wisnioski has described this movement as one with clearly articulated leftist agendas. Beginning in 1967, prominent researchers at MIT's Fluid Dynamics Laboratory shifted their work in part toward "socially oriented problems," including pollution control and desalination, in what was labeled a re-conversion of scientific research. The laboratory thrived under its new mandate, bringing in funding from federal and state sources and training dozens of graduate students in the new areas through the 1970s. Perhaps researchers at less established institutions doubted that their efforts at re-conversion would garner such support. MIT's good fortune may have seemed like either a token gesture by funders or perhaps a public relations gesture in a time of increasingly vocal public hostility toward conventional science and technology. But Chicago's engineering establishment had reason to see re-conversion as a movement with some potential for growth. A dramatic protest at the meeting of the American Academy for the Advancement of Science (AAAS), held in Chicago in 1970, advertised the willingness of established scientists and their students to retool in this way. This more radical expression of alternative politics in engineering arose as members of Scientists and Engineers for Social and Political Action joined with Science for the People, a student group founded at Harvard and MIT. Perhaps inspired by the protests at Chicago's Democratic Convention of 1968, the scientists and students proposed the overthrow of traditional sources of funding for technical research along with the "perverted" uses of scientific talents that such money had supported.[29]

Apparently, IIT's technical faculty did not join this movement making itself heard on their doorstep. The major proponents of the protest came from high status schools such as MIT, Harvard, Berkeley, and the University of Michigan. Perhaps these scholars felt more secure in their professional standing and thus able to challenge the system from which they derived income. Interestingly, IIT did create a Metropolitan Studies Center in the late 1960s as an interdisciplinary effort devoted to studying problems of the city including pollution, traffic, and educational and economic inequities. Engineering faculty taught core courses in that center's degree program, but as has long been the case in engineering and science, the daily conduct of technical disciplines proceeded at IIT largely apart from the goals or findings of interdisciplinary ideals. The general curriculum for civil engineering, for example, did not integrate issues of urban planning; that application was examined inside the dedicated center.[30]

There is no doubt that standards for promotion and tenure in higher education exerted pressures on academicians to choose areas of study that matched the priorities of their employers. As noted earlier, to have spent time on service activities (advising or tutoring students in need of remediation, for example) could be risky, since research and to a lesser degree teaching constituted a known basis for professional reward. Similarly, to expend effort on research projects that senior faculty and administrators found objectionable could put one's tenure and promotion at risk. Separating political and practical reasons for such objections is not easy. Any employer might question faculty involvement with an unfunded research topic. Even Bugliarello, a strong advocate of integrated social and technical study, recognized that "most research projects that emanate from the public sector have a strong interdisciplinary flavor." Therefore, he added, faculty involved with those projects would have difficulty "figuring out where to publish," thus hurting their chances for advancement.[31] But just as student admissions procedures express both practical and social priorities on the part of educators, so do standards for acceptable amounts of research funding or publication in one's academic discipline.

As sociologists have long recognized, professions do not dictate the values of their members in any consistent way, but people must find their employment in some sense to be commensurable with their values. At the same time, as Bugliarello's words capture, practitioners' understanding of professional opportunities informs their more global political choices: that is, choices about what kinds of social systems or social change they will advocate. Critics of technology in the 1960s and 1970s blamed industry leaders and politicians for some of the worst abuses of the democratic process in the United States: poverty, environmental degradation, unemployment, and urban decay. Yet industry leaders and politicians were the very people with whom a university engineering program had to get along if it hoped to sustain funding. Given this tension, engineering administrators and instructors faced extremely difficult choices as educated people acutely aware of the unrest around them and the patronage system above them.

The near impossibility of ascribing any simple set of values to those who directed engineering programs in this era is made clear in the case of UIC as it contended with Illinois' extremely powerful state political apparatus throughout the 1970s. The de-emphasis at UIC on socially informed technical research reflects a confluence of social relationships and knowledge-making practices, which together rendered academic attention to urban problems nearly impossible for UIC's engineering faculty.

Research on Urban Problems as Second-Tier Research

Within only five years of UIC's founding, new administrators responded to poor retention rates and disappointing research funding levels by upgrading the university's engineering programs (described in Chapter 4). It was at this juncture that the college raised admissions criteria for its students, and concerns about faculty hiring and retention took on a new urgency.[32] In a snowballing crisis, disenchanted engineering faculty left the school while others felt reluctant to come aboard the sinking ship, and research dollars went elsewhere with the departing researchers. Graduate study suffered accordingly. In 1970, Urbana was giving nearly 60 percent of its engineering degrees at the graduate level, with a full third of that campus' graduate degrees overall being granted in engineering. UIC, in contrast, struggled to grant a mere 6 percent of its engineering degrees on the graduate level. There is no question that engineering programs in the University of Illinois' downstate and upstate branches were not proportionately sized to the state's regional populations.[33]

UIC's faculty and administrators worked in constant awareness of accreditation pressures.[34] MIT might have been able to afford to maintain a laboratory committed to nontraditional areas of study, but UIC had not yet met minimal requirements in traditional areas and this made many administrators and faculty wary of departures from convention. Paul Chung, appointed dean of engineering in 1979, dismantled the unusual organization of UIC's engineering programs that had followed "functional" lines by offering majors in materials, energy, information, and systems engineering. These Bugliarello had held, reflected the analytical tasks undertaken by engineers. The college turned instead to familiar categories for its majors, such as civil engineering or mechanical engineering.[35] In many respects, UIC's original social welfare commitment to the community, evident in dozens of outreach projects more and less closely linked to academic programs, did not diminish. But the College of Engineering increasingly saw itself as having to function apart from such commitments. In 1979, a report on UIC's involvement with urban problems showed 104 "urban-related" courses offered through the College of Liberal Arts and Sciences (in addition to those offered through UIC's Jane Addams School of Social Work), but only eight of those courses were offered by the College of Engineering.[36] In most respects, the physical conditions and social distress of the city in which UIC sat remained remote from the work of technical faculties as researchers sought the funding and prestige associated with more conventional projects.

The disincentives that surrounded any loosening of academic/industrial links are writ large in the history of UIC's engineering college, which faced and rejected a mandate for such a loosening in the late 1960s. In 1961, the state's General Assembly had established the Illinois Board of Higher Education (IBHE)—a ten-member planning and advisory board to coordinate budgets and operations of all public and private institutions of higher learning in the state. Near the end of the decade, the Illinois legislature charged the board with creating a master plan that would incorporate the public university system as it expanded. The group, made up of academics and university administrators from schools around the state, endorsed what appears to be a socially progressive role for engineering education in Illinois. The IBHE urged UIC to train engineers in subjects that would specifically address urban problems: ecology, urban planning, the design of affordable housing, and mass transit. Reporting in 1970, the IBHE foresaw that "solutions to the problems of pollution, health care, and many other concerns of urban and rural society will require the continuing attention of the brightest and most highly educated engineers in increasing numbers." The board recommended the creation of graduate centers and institutes for environmental research and practices. These were to be comparable to the clinical settings provided in medical education, and to complement an expansion of engineering undergraduate curricula to include "engineering science" and "wider knowledge of the social sciences and humanities." But in terms of the major state university system, this mission, tellingly, was confined to the Chicago campus, rather than applied to the University of Illinois system as a whole. The board intended that Urbana retain its focus on conventional engineering subjects.[37]

UIC faculty and some administrators objected strongly to the IBHE recommendations, in part because the board's very existence seemed a threat to the school's autonomy. In responding to the IBHE, UIC's leaders also made it clear that if they were to fulfill the needs of Chicago's citizenry in the manner prescribed by the IBHE, it would "be at the expense of our obligations to the entire state.... Being a 'good neighbor' is one thing, but ... [ellipses in original]." Even though the IBHE's master plan explicitly mentioned expanding instruction in those technical areas noted above, it struck the university's leaders that the Chicago campus was being asked to serve urban issues and topics, and that that role "leaves out engineering."[38] Even at a land-grant school certain classes of applied research seemed to engineering faculty to be intellectually inappropriate, poised to undermine UIC's academic legitimacy. Many UIC administrators and faculty did not accept the notion that the school's land-grant legacy demanded research on city issues. As one observer of the controversy summarized: "Urban problems, unlike agricultural problems, are far more political ... often strategy problems, not knowledge problems." Many at UIC felt that "solving the problems of the city was not the university's core mission" and, urgent as those problems might be, there were "dozens of agencies of all sorts with that goal as their *raison d'etre*." As had previous generations of public servants, the UIC faculty supported the idea of the land-grant school as one which should exclude unprofitable pursuits: the best way reliably to serve society was to serve commerce.[39]

The separation of "knowledge" and "strategy" work put forth by UIC's engineers was not an inevitable one. It reflects professional opportunities as the engineers saw them in Chicago: decisions about urban infrastructure and resource distribution might well involve the products of technical work (roadways, mass transit, new energy systems, or housing stock), but technical experts could not expect to subsist by serving the demands of urban planners. A broader perspective might have suggested that the state take the place of existing commercial mechanisms and subsidize the design and development of new, socially progressive technologies. But Chicago's engineers understood that this was not an immediately practical suggestion. The IBHE tried to frame its proposed changes in attractive, practical terms for UIC's faculty. The board noted a dip in engineering employment at the time and pointed out that

> Engineering education can be improved mainly by shifting the applications' emphasis in anticipation of the ever changing needs of society.... What this means is that there is likely to be less need for engineers in the aerospace and defense-related industries and far greater demand for their talents in solving problems in such areas as ecology, housing, urban renewal, transportation, health care and the like.[40]

By looking more closely at the seemingly pragmatic stance of the IBHE, an agenda becomes apparent that corroborates UIC's worst fears about being fiscally and reputationally marginalized in the state's higher education system. From 1969 to 1971, the IBHE's executive director was James Holderman, a political scientist and vice-chancellor at UIC. Holderman was also the son of a powerful downstate Republican politician and a favorite of Republican Governor Richard Ogilvie. Holderman's advocacy of an urban mission for UIC enacted a subjugation of that campus to Urbana in resources and reputation, bolstering the position of the larger, older campus in the system. Holderman called for a focus at the Chicago campus on applied knowledge and associated that emphasis with a new sense of outreach and "concern for community." But he quickly qualified what such applications of knowledge would actually mean for the growth of UIC by citing a "tight" state economy and a national oversupply of PhDs that argued against comprehensive programming and extensive research there. His calls for "accountability" and "relevance" at the Chicago campus can be read against other statements he made supporting the severance of the upstate campus from the University of Illinois system altogether. Given the circumstances, UIC's engineers were probably justified in dreading the development of an urban focus at their institution, recognizing it as the institutionalization of their lower status in the prevailing knowledge economy.[41]

Holderman knew that the IBHE's proposed strategies and those of UIC faculty were at odds. He once described his position in Illinois higher education as situated "somewhere between the dog and the tree."[42] These words remind us of how layered progressive rhetoric can be, embodying personal agendas and local political concerns as well as, or instead of, broad social philosophies. Hopes for the development of socially informed engineering were by no means foreign to UIC's engineers. As noted, the unusual founding structure of the UIC's College of Engineering, based on engineering function rather than customary disciplinary divisions, probably arose from some progressive social strains in professional engineering that de-emphasized conventional measures of research productivity. George Bugliarello was actively interested in the emerging field of Science, Technology, and Society; while dean, he supported a history of technology elective course within the College of Engineering. Traditionally, the few courses that existed in this subject area offered descriptive time-lines of invention.[43] Bugliarello, by contrast, took a prescriptive approach to the history of technology. He imagined that exposure to the philosophy of technology might persuade engineers to move their research priorities away from market determinants toward collective societal benefits. Bugliarello fully believed that a thriving technology sector could be compatible with reduced industrial pollution, more sensible consumption habits, and even an end to U.S. military involvement in Southeast Asia.[44] But the integration of liberal ideology into engineering practice faced considerable obstacles in this academic setting.

The apparently progressive IBHE actually embodied a number of such obstacles in its differential plans for a traditional Urbana engineering school and the new "urban mission" engineering school it envisioned for UIC. That designation, without a vast commitment of resources, almost assured that the UIC College of Engineering would remain small and lacking in influence, with little hope of attracting accomplished faculty or bringing its graduates solid employment prospects. Given that the majority of funding, consulting, and employment opportunities in professional engineering did not encourage socially informed research in this period, any implementation of reformist ideologies in academic engineering would have required very strong institutional support. That kind of support for UIC was never part of Holderman's vision.

The research emphases preferred by UIC's engineering faculty reflected their understanding of institutional and professional opportunities, and in the case of their reactions to the IBHE's plan, their reluctance to lend power to a body they saw as unsupportive. Again, it is possible to see the choices made by engineering educators as pragmatic and thus as only indirectly ideological. For example, the financial costs of teaching are never negligible. Problem-centered engineering instruction is particularly expensive, requiring specialized facilities and equipment. If supported by generous grants or contracts as specialized industrial or military research might be, a school could readily undertake such pedagogical expenditures. But if problem-centered research and teaching is dependent on nonprofit or unreliable public funding sources, as investigations of mass transit or affordable housing would likely be, such work would be off-putting to all but the most well endowed schools. Neither UIC nor IIT could afford to see itself as regularly providing pro bono services. Similarly, if employers for engineering graduates were most numerous in traditional engineering fields, neither school would be likely to attract new students with a curriculum that stressed unusual fields such as urban environmental, housing, or transportation engineering. Many educators recognized that the interdisciplinary nature of classes on urban problems was itself an immense problem. Bugliarello certainly saw the pressure that industry exerted on content matters in engineering: "The requirements of potential employers . . . preserve traditional areas of study, such as Mechanical or Civil Engineering, even though this leads to duplication and awkwardness in the design of programs, as the boundaries between many of these areas have become very blurred."[45] Guidelines set by the professional engineer licensing examinations and accreditation processes reinforced not only the technical emphases of undergraduate curricula, but also strict disciplinarity.

Such constraints would discourage unconventional research and pedagogy in any engineering department hoping to succeed in the competitive academic world. Two questions remain that can only be answered with a broader cultural perspective: why did this set of practical conditions arise, leaving academic engineers with few outlets for their reform impulses; and how strong were those impulses? From within the academy, there seems on the surface to have been a great deal of support for technical work on urban problems. UIC's hiring of Dean Bugliarello and IIT's inauguration of its Metropolitan Studies Center are just two indications among many. The ACE's regional conferences in 1974 engaged dozens of deans, provosts, and college presidents in the question of how universities should serve urban America. The Ford Foundation funded the Harvard-MIT Joint Center for Urban Studies in the 1960s, with the support of MIT president Howard W. Johnson (who was also an early and vocal supporter of minority engineering education programs).[46] But academic engineering functioned within a shared national culture of engineering which grounded professional identity for technical experts in a realm devoid of (apparent) social intentionality. The obstacles to interdisciplinary work in academic engineering are in fact steeped in ideology, and arise from complex cultural disinclinations to social reform.

The Firewall

The idea that social and technical thought can be integrated in a single epistemological enterprise is counteracted at many levels by the norms of modern engineering. First, political disinterest has been a recurring byword of engineering in the modern era. As science has historically claimed a fundamental commitment to objectivity, so engineering has also identified itself with "enlightened impartiality" and ideals of "unfettered scholarly inquiry and response."[47] In a formalization process evident from the late nineteenth century onward, technical languages, quantification, instruments, and other forms of representation reinforced the idea that engineering work pared away discretion in favor of rational preordained techniques for decision making. A discourse of disinterest surrounded educational and professional practices. Quantification signaled procedural regularity, which in turn assured fairness. Through the twentieth century, engineering steadily instrumentalized measurement, design, and materials and structural testing tasks, which brought automation and calibration to bear on tasks which had once appeared to be far less regularized. On the shop floor or construction site, standards and specifications further helped regularize procedures and communications and adjudicate relations between technical practitioners, and between engineers and their nontechnical clientele. While many of these innovations disguised, rather than eliminated, subjective judgments in engineering work, they nonetheless supported a value-free aura for engineering.[48] We can presume that this discourse was offered in good faith, not as obfuscation. In one much studied case of the 1970s, for example, American highway engineers reacted to community protests with great surprise and dismay. Activists accused designers of the interstate highway systems, which often cut through inner city neighborhoods, of creating structures that privileged affluent white, suburban commuters and isolated poorer minority citizens in degraded settings. In at least one instance (in Boston), engineers responded that they had been following specifications provided to them by government planners, and were genuinely confounded by their involvement in the controversy. Their work practices, based on following highly codified sets of technical instructions, induced a sense of remoteness from politically freighted decision making.[49]

It would be a mistake to assume from such examples that training and professional habits rendered all engineers devoid of political vision. In a second current of engineering self-identity, claims of precision and quantitative rigor sometimes justified engineers' involvement in matters of human welfare. That role, however, was based on the epistemic distinctiveness of engineering work from other means of social interventions. Even when Bugliarello and his collaborators on the 1973 "Impact of the University" report associated the quantitative slant of engineering expertise with an important element of policy making, the nature of problem-solving techniques remained distinct from humanities and social science methods: "If there is one thing that all social problems have in common, it is that they have many variables and engineering schools are the only ones producing students ... that are not afraid of five or more variables. In fact, their interest is proportional to the number of variables involved in the problem."[50]

This self-concept for engineering does not seem to differ significantly from arguments of the Progressive Era, when engineers articulated a technocratic approach to the identification and solution of social problems. To those earlier practitioners, the quantitative features of engineering rendered it superior to such fields as politics and architecture when it came to upgrading urban infrastructures and dictating healthy lifestyles. The differences between engineering and more subjective disciplines made engineering a good candidate for the planning of reforms. More common in the 1960s, however, were approaches to social or policy problems that simply disassociated technical and social or political activities. David Hammond, the BART engineer who had recommended engineers' involvement in humanistic aspects of public works, nonetheless contrasted public transportation projects with "space shots and sophisticated computer networks." He saw the mass transit system as solving "basic people and political problems, not technical or engineering projects." This delineation is similar to the UIC engineers' concern that attending to Chicago's urban problems would "leave out engineering." What is more, Hammond saw difficulties encountered in the building of BART as the fault of "people and political" forces, as he put it, exerting an unfortunate influence on engineering. Questions of blame aside, he is positing here two completely separate areas of intellectual and institutional endeavor.[51]

Discussions of engineering education and labor markets supported such distinctions. We can historicize the fact that Western cultures differentiate between "social" and "economic" factors in the formation of public life.[52] The arbitrariness and political implications of that differentiation for the case at hand have recently been articulated by historians of urban planning who explain how categories such as race, gender, class, and sexuality are "socially produced by practices such as zoning, infrastructure, housing and transportation."[53] When Chicago planners studied "Conditions and Trends" in the city in 1967, they divided their report into studies of population (including education and migration trends), economy (including businesses and taxation patterns), and land (tracking real estate values, housing, and commercial development). These familiar categories of analysis helped to naturalize the idea that economic activity can occur apart from social considerations.[54] To suggest that social matters can or should be *introduced into* economic decision making obscures the fact that economic priorities are born of social priorities. To obscure that connection is to reduce the likelihood that profound changes will result in either realm.

Significantly, this kind of boundary work was an ongoing project for engineers after World War II because the war ended any simple equation of quantification with material tasks or outcomes. Math and even physics evidently could yield knowledge about social matters, as well. The expansion of wartime operations research (OR) in the late 1940s meant that techniques once applied to the analysis of complex but extremely concrete matters (such as gun accuracy or the mechanical reliability of weapons) began to incorporate the study of social phenomena (strategic decision making in combat, or inter-service cooperation). The founding of the Air Force's Project RAND as a nonprofit think tank in 1948 brought powerful new computing technologies to bear on systems research, and behaviors that once might have seemed eligible only for psychological or sociological interpretation were now subject to mathematical or physics-based analysis.[55] Management science became a fruitful peacetime subject for such experts. Economists and political scientists at RAND quickly found an audience for quantified models of market and electoral behaviors; the rise of Rational Choice theory in the 1950s meant that inchoate qualities such as "power" or "influence" could be modeled, and thus, the Rational Choice theorists promised, their emergence and impacts predicted.[56] Medicine and social science had long used numbers in the service of social goals, devising instruments (both mechanical and mathematical) as needed. But now engineering faced dramatic evidence that in the highest-tech contexts of military-industrial research and development, science and politics were smoothly blending. Jennifer Light's "defense intellectuals" even found their way into the study of urban renewal through these channels. What did it mean for engineers to insist in this period that "technical" and "social" were two different epistemic categories?

The compartmentalization of technical and social thought could convey a sense of the neutrality of professional practice. The universalizing techniques of science bring credibility to the idea that research in science and technology proceed in a fashion detached from social conditions.[57] The adoption of an apolitical posture for engineering took on greater significance as campus unrest increased through the 1960s. John Rettaliata, then president of IIT, was appalled at student requests that he issue a statement condemning the Vietnam War in 1970. He defended his reluctance on the grounds that a tax-exempt institution such as IIT "has no business getting involved in politics."[58] Given the prevailing conditions of campus unrest and a generationally polarized nation, he can hardly have presumed that universities might easily achieve a lofty remove from societal troubles. Some politicians, for their part, outlined "political problems" as distinct from "knowledge problems" when they wished to deny their own responsibility for social conditions. In a reversal of the dynamic that Hammond described, wherein BART's engineers suffered the whims of political interests, Senator Daniel Patrick Moynihan of New York, writing in 1970, saw the "other" sphere (that is, knowledge making) as having more agency than his own. Writing about the relationship between knowledge problems and political problems, he wrote: "There is an increasingly dense mutuality between the two, it being a defining characteristic of technological society that what can be done must be done, so that where knowledge comes into being on its own, as it were, it commonly produces pressures within the political system to adopt the social goals made possible by such knowledge."[59]

Moynihan envisions a society in thrall to science and technology, in which politics cannot conduct itself freely. In all such rhetoric, voices situated in separate spheres—government, industry, and academia—concerned about social change pointed to dimensions of culture outside their own control that could serve as evidence of their own blamelessness.

As demands for social accountability for science and technology grew throughout the postwar decades, liberal factions challenged that separation. They identified the inherently political character of knowledge and the demonstrable social connections among those spheres. Such connections prompted Eisenhower's identification of a military/industrial complex. Meanwhile, representatives of technical occupations defended their actions with rhetoric that often reinforced their elite self-identity and the distinctive character of scientific knowledge. As early as 1952, Rettaliata answered complaints about science (which had been voiced "even before the shadow of the atomic bomb darkened civilization's sky") with a rebuttal painted in broad strokes: "Science is feared. Man's future is doubted. Among large numbers, science and its effects are profoundly misunderstood. Scientific illiteracy is as dangerous in the great body of the people as economic or any other form of illiteracy."[60]

Over the following decade and a half that fear and doubt about science solidified on many college campuses, and undergraduate enrollments in scientific and technical disciplines dropped in what many educators saw as a direct consequence of that public crisis of faith in science. John Kemper, dean of engineering at the University of California, Davis, wrote in 1973 that "young people have been presented with some pretty bad criticisms about engineering professions."[61] The authors of the report on the impact of Illinois' engineering colleges believed that: "the demand for engineers and scientists has changed from 1968, when it was partially unfulfilled, to 1972 when the concentration on urban and social problems and the highlighting of threats inherent in pollution engendered a falling popularity of engineers and scientists."[62]

Exposure to the university itself, the historic home of scientific knowledge, now began to alienate young people from science. The ACE reported precipitously dropping interest in science among college students from the time they entered college in 1967 to the time they left in 1971, with their interest in English, psychology, sociology, and anthropology increasing markedly.[63]

Schools like IIT, having based their public identity on their technological prowess, had to carve a new identity that accommodated these cultural anxieties while bolstering the status of engineering. One 1965 newspaper editorial defined "Chicago's stake" in the school as virtually life or death.

Endorsing IIT's new campaign to raise $25 million in private donations and public funding, the *Chicago Daily News* told readers that the nuclear age calls for "leaders in whom scientific and technical know-how are complemented by human understanding." The city also displayed far more immediate, if less universally compelling, problems in its impoverished and racially fraught central district, and all institutions of higher learning needed to chart a course that both acknowledged and controlled public discomforts with so-called establishment interests, now being blamed for many problems.[64]

The Self and the Collective

These tensions about culture and public welfare tapped into long-standing debates about the role of universities in the United States. There is an old argument in American higher education that associates humanities with individual self-development while casting scientific and technological fields as collective enterprises destined to modernize entire polities. This distinction has served engineering in different ways at different historical moments. Through the mid-twentieth century, the personal and civic functions of education meshed well for technical disciplines: American universities articulated the role of humanities education in the maturation of students into fully rounded individuals. For several generations, the provision of humanities classes to engineering students promised to instill both moral rectitude and professional aptitude, and thus contributed to the popular definition of college educated engineers as good candidates for managerial and civic leadership roles.[65] As voices critical of science and technology became louder at mid-century, academics began to address those critics with rhetoric defending the inherent altruism of science itself. Rettaliata bluntly took on early, Cold War versions of some of those critiques: "Social inequality has marked the pages of all human history. Struggle between the haves and the have-nots has scored all ages. . . . It has always seemed to me that socialism consists essentially of a leveling *down* process. Science, technology and production are showing us how all men may be leveled *up*."[66]

Science was still being defended in these terms in IIT's promotional rhetoric of 1974. Acknowledging in one phrase both the dangers and promises of its work, IIT noted that universities centered on science and technology now faced "heightened responsibilities." Specifying exactly where it fit in the discomfited city, the school said, "IIT's curriculum has combined technological disciplines, so important to the modern urban society, with humanistic disciplines, so important to the individual."[67] This kind of formulation maintained the century-old model of the culturally sophisticated engineer who could lead society forward, and claimed that one practitioner could embody and reconcile the two cultures even in an era that doubted the humanism of technology. With few exceptions, however, the actual daily work of engineers at IIT did not take that reconciliation to the bench level. Coursework in technology and humanities areas remained distinct, with skill sets for engineering students adhering to disciplinary convention.[68]

Bugliarello wrote and edited publications on the philosophy of technology; that part of his work represents a particularly ambitious attempt by an engineering educator to understand the social origins and impacts of technology. As noted, he had profound doubts about the applications to which technology was being put in the later twentieth century. He believed that the university could play a role in changing that situation. However, in 1971, in a version of a talk he had given before the AAAS earlier that year, Bugliarello identified problems preventing that kind of reform. First, he understood the university to be mired in the priorities of the market. Traditionally, Bugliarello wrote, in following the narrow mandates of industrial productivity, the university holds to conventional curricula and a discipline-based organization that hamper its ability to detect and correct major social disjunctions. Simply overcoming traditional divisions between academic fields would not ensure reform because neither optimistic scientists nor pessimistic humanistic "protestors" could see that in a healthy society, technology must "integrate man, society and machines."[69]

His cultural critique is remarkably nuanced in places and anticipates later developments in the sociology of scientific knowledge. For example, in this same essay Bugliarello describes the My Lai massacre in Vietnam as a product of bureaucratic dependence on quantification. Needing a metric by which to determine officers' eligibility for promotion, the military relied on body counts. Maximizing enemy deaths became a "norm" for all those functioning within that knowledge economy. For Bugliarello, this kind of narrow instrumentality configured the work of virtually everyone who devised and deployed technology in the 1970s. The efficacy of the measurement tool obscured its human effects as it sustained institutional habits. Some twenty years later, Theodore Porter's influential work described the historic reliance of engineers on cost-benefit analysis and other quantification techniques, identifying just such an intersection of professional and epistemological strategies.[70]

Some other engineering educators of the day understood that cultural expectations and institutional structures were inseparable; the behavior of academics, seen as a social modality, reflected both. A 1968 committee of the American Society for Engineering Education (ASEE), studying possibilities for collaboration between engineers and social scientists, saw in this linkage a conundrum: a shared intellectual base "which substantive communication might well help construct, needs to be present to some extent for substantive communication to take place." Substantive communication, ASEE committee members recognized, is dependent on what we would today label participant "buy in." A collaboration works best, they wrote,

> ... when it is concerned with a specific problem or task which falls within the sphere of interest of both participants. When one is merely helping the other solve his problem, the relation is different. An engineer is interested in the technical aspects of a pollution problem. He recognizes that it has certain social aspects as well, and he calls on a social scientist for advice. No new joint field of professional interest is developed, and no real foundation is laid for further communication.[71]

Rewards for collaborative work across departmental or school lines are also vital, the writers note. In these observations they further articulate social structures of opportunity, credit, and authority to which Bugliarello had pointed.

In the introduction to their 1979 volume on the history and philosophy of technology, Bugliarello and fellow UIC engineer Dean B. Doner call for concerted attention to technology's role in the perpetuation of war and poverty. They pose the question of "whether technology itself carries values or creates them." Yet they also declare technical "discovery" to be "without method" and "[occurring] outside priorities."[72] To ascribe a "serendipitous" character to technical investigation is to miss the mutually shaping nature of organizations (such as universities or research and development laboratories) and social values. It is also to miss the kinds of tactical decisions that technical practitioners and educators must make on a daily basis (like the officers at My Lai to whom Bugliarello referred) which might lead to the perpetuation of certain, socially problematic technologies. Thus, a certain degree of technological determinism comes to underpin their explanations of contemporary culture: "The development of rapid technological innovation, imposing adjustments on people faster than they can absorb them, breaks choice down. When technological innovations break too rapidly upon man, he can order his technological priorities only in theory; in fact he simply does not have enough time."[73] Nonetheless, the sense of momentum the authors identify does suggest the problem with severing processes of invention from their societal origins.

The identity that engineering educators sought for their field thus enlisted humanistic thought as an appropriate and laudable feature of modern technical practice. But educators generally promoted humanistic concern within engineering at the level of identity formation, as a sign of a compassionate or sophisticated worldview on the part of students, not on the level of problem choice in the classroom, laboratory, or place of employment after graduation. The larger scientific establishment insured that the firewall between societal and scientific thought would not weaken. As was the case with standards for student admissions and performance described earlier, standards for technical excellence (that is, definitions of valuable engineering work) supported that politically conservative approach to technical work. In 1974, the NSF described its approach toward projects intended by universities to bring more women and minorities into science and engineering through recruitment and improved retention. The agency labeled such projects as ineligible for funding because they were "crash programs," not research projects designed to produce "sustained" contributions to the nation. This declaration did the work of maintaining inequities in two ways. First, it simply denied monetary support to diversity projects. Second, it confirmed a disconnect between social reform and significant intellectual endeavor. It did not do so on the basis of any political claim, which might be subject to political criticism, or through the simple but obviously conservative argument that diversity projects were not within the purview of the NSF. Rather, it differentiated social from scientific activity on the basis of likely historic impact: the former was identified as having only short-term impacts, the latter as promising long-term benefits. It is a somewhat circular argument, but all the more unassailable for that circularity.[74]

Conclusions

Distinctions among engineering subdisciplines reified the distance between technical and social agendas. As engineering fields moved increasingly toward narrow specializations, graduate programs increased in numbers. Undergraduate programs, in turn, implemented rigorous specialized courses to prepare students for graduate training. This trend helps to explain why UIC's unusual organization of engineering departments by function collapsed after only a few years. There is a broader explanation for that collapse, as well. As Thomas Gieryn and others have pointed out, it is only by limiting membership that groups can hope to reach consensus among their members. The process of professionalization (expressed in engineering through such functions as funding, hiring, and tenure decisions) defines both eligibility for membership in a profession and the content of members' work. In science and engineering, as in every other profession, agreement about the nature of best practices reflects shared ideas about how professional status might best be achieved. That is, standards for fundable engineering research that demanded only minimal attention to social issues remained hegemonic because so few educators chose to challenge those standards. Standards for accrediting curricula that minimized work in urban problems discouraged universities from breaching those standards. Employers who were eager to hire engineers with conventional skills provided no incentives for universities to teach unconventional courses. In the "Regional Role" report, Bugliarello and his coauthors found that it was "peer-group attitudes," rather than "administrative policies," that were holding back engineering faculty's involvement with public service projects. Even in the absence of explicit policies against public service, engineering faculty were likely to follow research paths that ensured the greatest institutional support.

This depiction of engineers' choices, as educators and researchers functioning in an uncertain political climate, leads to several questions. How do we go from that set of rational decisions to an understanding of value formation? What is the connection between conservative research agendas and non-democratic features of the university? A return to the general cultural climate in America may help us answer these questions. Amid a great deal of negative sentiment toward technology in 1970, and a shrinking employment outlook for engineers as aerospace and defense industries contracted, one observer held an optimistic view of engineers' employment prospects. In a report for the Engineering Manpower Commission, employment and accreditation expert John Alden wrote: "The basic demand for engineers and technicians in the U.S. employment market stems from the steadily growing production and consumption of goods and services. Not only do more people want more things, but the things they want are becoming increasingly more complex."[75]

What was good for business was good for engineers, obviously, and business rationale consistently pointed toward research and development that supported material expansion. Academics came to be very comfortable with that set of priorities. Christopher Newfield, building on research by social scientists Barbara and John Ehrenreich, depicts university faculty as an emerging "professional managerial class" sympathetic to and well-served by the goals of business owners and leaders, and as experiencing a sense of "deep comfort with non-democratic group life" in the later decades of the twentieth century.[76]

While most academic disciplines after 1950 weighed research more heavily than teaching when distributing resources to individuals and programs, engineering carried the additional burden of an assumed economic role. That role made high-tech contributions to the military sector and private industry a high priority for university engineering departments. It also encouraged the pursuit of time- and cost-saving applications of engineering to civil projects (which often employ products and services that bring profits to industry). Historic ties between academia and the military-industrial sphere had intensified through the 1950s, just as IIT and UIC leaders began planning their modernized engineering curricula. Substantial shifts occurred in engineering departments during this decade. As an "avalanche" of federal money became available for research, principally from the military and the Atomic Energy Commission, American universities moved from undergraduate emphases on technical skills such as machining, surveying, or drawing and toward more rigorous courses centered on scientific and mathematical subjects. Graduate engineering programs, including an unprecedented number of doctoral programs, expanded to serve short-term research agendas within universities and longer-term industrial and state interests in scientifically informed engineering research. Productivity, in products or knowledge and personnel for industry, became ever more important and easily measured for academic engineering departments.[77]

The immediate outcomes of this emphasis on productivity included enhanced status and material gains for individual faculty members, departments, and universities, as well as for the nation's economy. Within the terms set by economic indicators, it made sense for engineering schools to turn away from the address of social problems and from any curricular programming that might dilute a "cutting edge" reputation. As we have seen, remedial coursework or altered timeframes for undergraduate degrees that might have led to greater minority inclusion in Chicago fell under this heading. If we cease to understand growth as the first or only cause of American economic decisions, we can begin to see that a hierarchical society also justified these exclusionary priorities in research and teaching. By the 1980s and 1990s, a conservative turn in many parts of the country made that agenda particularly clear. Assertions of reverse bias in which white students were seen to have sacrificed university places to less accomplished minority applicants became more common, and legal reversals to affirmative action policies increased. Within these legalistic campaigns, merit served an important function. The idea of a value-neutral index of student talent explicitly supported arguments against race-based or other compensatory educational policies. Critical inquiry into the vitality of

America's democratic social structures became newly unfashionable among politicians, and in the academy, many gains made during the civil rights era faded as the nation's leaders pushed for "color-blind" sensibilities for the new millennium. Race as a topic of address did not disappear altogether. A new understanding of diversity, favorable to the interests of globalizing corporations, developed, heralding a kind of pluralism that would not threaten established social stratifications. The firewall that had previously kept the address of social matters from shaping technical practice was replaced by something more akin to a mask, in which multicultural expressions of tolerance disguised the conservative ideologies of global capital. It is these discouraging trends, and some more hopeful progressive exceptions in engineering research and pedagogy as the century turned, that I next consider.

CHAPTER SIX

Race and the New Meritocracy

*Engineering Education in the Texas A&M System,
1980 to the Present*

PRAIRIE VIEW AGRICULTURAL & MECHANICAL UNIVERSITY, the historically black branch of the Texas A&M System, rises from a flat expanse of plain about an hour's drive northeast of the system's much larger, predominantly white main campus in College Station. In 1977, Alvin I. Thomas, then president of Prairie View A&M, offered his students advice as they prepared to enter the world of work, many in professional fields that were newly opening to African Americans. When the graduates encountered challenges along their career paths, Thomas urged, they should distinguish between obstacles presented by "discrimination" and those that arose as a result of "competition." The former should be deemed a regrettable vestige of racial bias; the latter, he explained, represents a consummate feature of the democratic American marketplace.[1]

Thomas clearly meant his advice to be encouraging, providing students with a constructive approach toward professional adversity. But in some ways his outlook was utopian. First, it mistakenly imagined that U.S. citizens of color could now hope to encounter with regularity the vaunted "even playing field" where true equality of opportunity had replaced discrimination. Thomas had some reasons to be optimistic, but countervailing tendencies were clear. The legal and societal reforms of the civil rights era that Thomas had witnessed had, of course, included new hiring regulations, affirmative action programs, minority set-asides, and other broad correctives to bias in education and employment. In fact, as Thomas spoke, progressive Texas politicians were laying the groundwork for an unprecedented redistribution of state higher education funding to include historically black public universities. Founded as land-grant schools in the 1870s, Prairie View A&M University (PVAMU) and Texas A&M University at College Station (TAMU), had embodied prevailing ideas of racial separation in the South for over one hundred years. Most Texas institutions of higher education were officially integrated by 1964, but TAMU remained often indifferent and even inhospitable to racial diversification. It offered professional mobility to many white Texas families, contributing large numbers of its graduates to military and technology sectors, but few black or Hispanic Texans attended the school as informal discrimination in admissions and a lack of targeted recruiting and scholarships kept old patterns in place. PVAMU remained, throughout these decades, drastically underfunded. Now, finally, as the 1980s approached, pressure from the federal Office of Civil Rights (OCR) and the hard work of regional political supporters of PVAMU culminated in changes to the state constitution and the first flow of serious public funding to the historically black campus.[2]

At the same moment, however, the nation was on the verge of a conservative retrenchment that negated many promises of the civil rights era. By the 1990s, in many parts of the country the very idea of competition had become synonymous with a rejection of governmental intervention in economic processes. Educational opportunities for minority citizens stagnated or diminished in many disciplines, including engineering, by the century's end. In the year 2000, twenty-three years after Thomas's speech to his graduates, the OCR determined that PVAMU remained an inadequately funded sibling to its massive, historically white counterpart.[3]

Moreover, the line between occupational "discrimination" and "competition" that Thomas posited has always existed in some disciplines as only a very blurred distinction in the United States; the categories themselves may not be nearly as separable as Thomas implied. As we have seen in the previous chapters, notions of intellectual merit in engineering arise from historically contingent presumptions about rigor, selectivity, and productivity. In U.S. engineering, what has counted as talent and authentic competitive advantage has always had a great deal to do with structures of social opportunity, not just the behaviors or abilities of individuals. Since 1980, critics of compensatory educational policies in general and of affirmative action in particular have argued that the maintenance of high educational standards and national economic performance depend on a race-blind approach. This argument set the stage for retrenchments in minority science and engineering education and for limits to the occupational opportunities on which Thomas seemed to be counting. Contrary to his expectations, in the eyes of some socially progressive and very worried citizens, the conventional imperatives of competition had, by the 1990s, become a means of discrimination.[4]

The approach taken toward racial diversification by the Texas A&M system's engineering programs at Prairie View and College Station in this increasingly conservative era offer the final case studies in this book. Examining this latest episode in higher technical education over the end of the twentieth century and opening of the twenty-first, we come across some social phenomena that should by now be familiar from our consideration of race in U.S. engineering education. For example, in 1980 the development and dissemination of new technical knowledge continued to hold for many the promise of both individual and national economic uplift. In addition, nearly two decades after the Civil Rights Act, the terms of debate in educational and policy circles continued to center on apparent conflicts between the pursuit of "inclusion" and "excellence," two distinct analytic categories for many North Americans who still saw the historical underrepresentation of minorities in science and engineering as a result of differential abilities, not of retrograde cultural custom. Fears for the preservation of U.S. productivity and international security still loomed over, and even justified, discussions about equity in higher education, as they had since mid-century. Conservative educational policy now routinely and, one might argue, disingenuously given the unfolding of the civil rights era, denied the significance of practitioner race or gender and supported the idea that technical knowledge did not have social features or functions.

That denial had serious consequences for African Americans' entry into engineering. Black enrollment in U.S. science and engineering doctoral programs dropped 20 percent between 1996 and 1997 as Texas and California passed laws constraining consideration of race in higher education.[5] Conservative voices reasserted the association of good scientific practice in classroom and laboratory with the erasure of ethnicity, and in many cases, of gender as well. Federal and state educational policies perpetuated that erasure through a return to race-blind performance standards: seemingly value-neutral measures by which young people might be admitted to university engineering programs and by which the success of those programs might be judged. Crucially, diversity did not disappear as a priority for education policy makers. In the mid-1990s, when expansionist goals of Western business interests began to make an ethnically diverse corps of managers seem vital for the conduct of international commerce, workplace diversity gained a new appeal for many corporate employers. University science and engineering programs, eager to fulfill the labor needs of employers eyeing international markets, supported that goal, and a large part of the funding for minority engineering programs in the late twentieth century arose from such business sources. Yet this corporate application of diversity only selectively addressed the historical underrepresentation of African Americans in technical occupations. We must consider, as in our earlier cases, exactly what counted in the United States as sufficient race reform after 1980 as not only Americans of color, but citizens of many other nations experienced the negative effects of U.S. racialist thinking in an increasingly globalized economy.[6]

Texas serves as an excellent case study with which to probe these tensions between agendas of social change and stasis. This southern locale only slowly relinquished its segregationist legacy. It did not officially repeal language in its state constitution permitting separate but equal schools until 1969.[7] For some observers, older populist currents and large minority populations historically have held out the promise of a liberalization of Texas politics. For example, one widely followed post-World War II theory held that as Texas energetically industrialized in the later twentieth century, working-class citizens of all races would unite against oppressive management factions. In so doing they would dismantle the quotidian bases of racial conflict. This, of course, did not happen, but as historian Amilcar Shabazz found, members of a strongly organized black professional-managerial class did take on the task of integrating public education in Texas. Their successful efforts reveal mechanisms by which even some of the nation's most intractable racist legal structures might be challenged.[8]

To the present day Texas remains a site of strongly held beliefs regarding race on both ends of the political spectrum. From the standpoint of higher education, Texas' political record reveals powerful efforts against affirmative action, most notably the 1996 decision of the U.S. Court of Appeals Fifth Circuit in *Hopwood v. Texas*. This decision, arising from a suit brought by white students denied admission to the University of Texas law school, prohibited consideration of race in the admissions of all universities located in Texas, Louisiana, and Mississippi. Enrollments of black and Hispanic students in both private and public universities dropped precipitously after *Hopwood* and subsequent so-called race-blind alternative plans for minority inclusion in science, engineering, and math failed to correct this decrease. Nor did the U.S. Supreme Court's partial reversals of this ruling in 2003 fully revive affirmative-action efforts in the Texas A&M System, as leadership at the flagship College Station branch decided anew to reject racial considerations in admissions. Black-identified educational institutions faced new obstacles in this atmosphere, at great cost to the diversification of engineering.[9] And yet, since that turning point in the early 1980s when the federal government first insisted on new efforts to integrate Texas universities, some educators at PVAMU and TAMU have achieved small but continuing success in their efforts to draw minorities into engineering and thus into high-tech military, industrial, and academic employment. At PVAMU, a long-held mission of broadly inclusive admissions segued in the 1980s to a more selective form that nonetheless maintained an emphasis on the school's black heritage. Dramatic upgrading and expansion of the school's science and engineering programs were central to that transformation. TAMU, for its part, has been innovative in the area of minority engineering programs (MEPs) since the 1970s. Both schools have drawn minority students into the highest levels of engineering research circles, making inroads in the persistent problem of graduate and faculty diversity.

The simultaneous resistance to and embrace of social reforms in engineering education—seemingly contradictory developments—at PVAMU and TAMU are followed in this and the next chapter. Once again questions are raised about why efforts put toward the correction of occupational inequities achieved the form and scale they did; why was *this* much money provided for higher education diversity efforts, and not more? Why the provision of *some* programs or scholarships, and not others? The nation's gradual but marked withdrawal from identity politics forms the backdrop of this narrative. To many Americans in the late twentieth century, pronounced expressions of ethnic identity seemed at odds with mainstream social ideology, let alone with the rarified intellectual activity commonly associated with scientific achievement. Race activism based on collectivist ideologies contradicted some of the basic tenets of corporate culture, and like other liberatory movements of the previous decades now moved away from the center of public awareness.[10]

Historically black colleges and universities like PVAMU had long seen their purpose as "not education for its own sake but for a continued challenge of the status quo in oppression, race relations and social behaviors."[11] The chapter tracks PVAMU's approach to that dual mission. By 1990, simply by systematically including minority Americans, HBCUs were raising the question of what does and does not count as science: Under what conditions can the *social* also be the *technical*? Can the reform of occupational opportunity structures (such as the racial diversification of engineering) be part of legitimate knowledge making (here, engineering teaching and research) in post-Reagan America? In Chapter 7 I provide a description of MEPs at TAMU that were created in response to federal pressures after the mid-1970s, in which diversity initiatives led to curricular changes in that school's College of Engineering. These changes in some ways reasserted and in others boldly departed from priorities of the MEPs developed during previous decades. Their history highlights the variable levels of radicality with which educators have approached the problem of exclusion since 1980. In addition, high-tech research undertaken at PVAMU, for which NASA supplied funding as part of its program for the support of research at HBCUs, challenged fundamental concepts of what constitutes fundable engineering research. This research offers an example of one way in which race-conscious science policy may produce both valuable science and redistributions of economic opportunity.[12]

Engineering at Prairie View: Modernity and Opportunity

In 1876, higher education in the town of Prairie View began with the establishment of Alta Vista Agricultural College, the state's answer to the Texas constitutional requirement of a school for "colored youth." The college, administered by the Texas A&M Board of Directors, opened with a budget of $20,000 and eight pupils. Whether these young men picked cotton as part of their curriculum, as one history of the school maintains, is not clear, but there is no question that the school held much the same status as Maryland's black public institutions of the period. From its inception, the college represented a means by which white authorities could acknowledge but severely limit the rights of minority Texans to public higher education.[13] As Texas A&M at College Station and the University of Texas (centered in Austin) in this period educated the state's white population with an ambitious range of liberal arts and scientific programs, the small school in Prairie View functioned as a place holder for the project of black self-improvement. The campus fulfilled mandates of the federal Morrill Act and a provision in the Texas constitution calling for the creation of separate schools for black and white citizens; within a very narrow interpretation of land-grant ideology applied to black campuses, this meant the provision of agricultural and industrial training.

In 1879, the small school became the Prairie View Normal and Industrial College and began to train African American teachers. As Shabazz has recounted in his history of PVAMU's first century, by the end of the 1940s the college had 2,400 full-time students and had produced 70 percent of the black teachers then working in Texas, but planned to offer new courses in law, engineering, medicine, pharmacy, and journalism. The NAACP had by this point begun directing its attention to upgrading black higher education in many border and Southern states. In Texas, the organization's efforts led to the creation of Texas State University for Negroes in Houston (now Texas Southern University [TSU]) and supporters of Prairie View College saw reason to hope for an expanded set of offerings at their institution.[14] Through the 1950s, the construction of new dairies, gardens, and chicken hatcheries reflected the college's ongoing commitment to agricultural training, but a seemingly more complete university model emerged as well. New divisions of Arts and Sciences, Nursing, and Engineering were established, as well as expanded offerings in Agriculture and Home Economics. In 1950, the NAACP achieved a U.S. Supreme Court ruling in favor of admitting Heman Sweatt, a black NAACP activist, to the all-white law school of the University of Texas. Despite strong segregationist governance in the state, *de jure* integration reached Texas by the mid-1960s and many advocates for the college at Prairie View began to hope for more reliable support for its development.

Paired goals emerged among PVAMU's supporters as integration of higher education gained national footing.[15] The school aimed to become integral to the state's economy, in the tradition of majority land-grant universities around the nation. It also sought to prepare its graduates to compete at many levels of the labor market that had historically been out of reach for black Texans. As the nation gradually brought its attention to the problems of urban poverty, Houston's large minority population hoped to find expanded opportunities for nearby training in a range of occupations.[16] The college's leaders wanted to move beyond a "provincialism" they associated with a primarily practical curriculum. Furthermore, when more liberal states began opening their college admissions to African Americans through the civil rights period, many historically under endowed HBCUs feared a loss of their best applicants to those more congenial settings and sought to upgrade their academic status. In 1966, PVAMU installed Thomas, originally an instructor and then dean of Industrial Education at the school, as president to help it contend with these challenges.[17]

In seeking fuller participation in the Texas economy and in employment opportunities, PVAMU gave precedence to the development of engineering and other technical fields it perceived as central to the state's development. It still accommodated the needs of the agricultural and lower-tech trade or mechanical sectors which had traditionally employed its graduates, but resisted relegation of its technical work to those categories of expertise alone, as might have followed from its historically subordinate status. If hiring opportunities did open to minorities as educators hoped, PVAMU's potential audiences in regional industries would be large. From about 1900 onward, Texas had seen steady expansion of its cities and industrial sectors as agricultural and cattle holdings gave way to growth in manufacturing and transport. Civil engineers had helped to implement the dramatic expansion of Texas ports and roadways, but it was mechanical and chemical engineering that found nearly limitless demand in the petroleum industries. Following World War II, Houston became the center of a multibillion dollar petrochemical industry, and in the second half of the century the state's historical identity as a cotton, cow, and oil economy shifted decisively to higher-tech areas.[18] With little historical representation in durable goods industries, Texas' involvement in aerospace engineering, inaugurated with the 1962 opening of NASA's Manned Spacecraft Center (later called the Johnson Space Center) near Houston, carried particular importance for the state's leadership. Engineering specializations in metrology and computation originally developed to serve the oil industry gave rise to Texas Instruments, carrying the state to national dominance in electronics and computing. In 1959, Texas ranked eighth in the nation in electronics manufacturing, but by 1980 it had risen to third, falling behind only California and Massachusetts. Computer-systems design has grown consistently since 1990, and as energy-related industries waned in the last two decades of the century, the rise in electronics and other high-tech industries in Texas have held steady.[19]

The state's public universities were integral to these developments. In the tradition of U.S. land-grant schools, the Texas A&M system established the Texas Engineering Experiment Station in 1913, a research entity that remains immensely influential today.[20] In addition to TAMU's status as a land-grant institution, the federal government designated the school as a "sea-grant" institution in 1971, and a "space-grant" recipient in 1987, acknowledging the wide-ranging role of the university across emerging technical sectors. Regardless, strong Southern resistance to the notion of a racially integrated society persisted even in rapidly modernizing economic and commercial arenas. Historically, the scale and scope of engineering programs at Prairie View had reflected the school's secondary status in the state system, a ranking particularly clear in the tensions between vocational and professional aspirations for technical fields at the school. From early coursework offered only in very narrow areas, such as classes in power plant engineering taught in the 1890s, PVAMU gradually built up its engineering programs until beginning in 1932 four-year courses were offered through the Division of Mechanical Arts.[21]

In 1937, PVAMU professor of sociology Henry Allen Bullock identified a trend in the labor market that reflected "the shift from a handicraft to a machine economy." Shabazz summarizes black educators' prime concern in this period as "closing the gap between the imperatives of modernity and the status of black education." In keeping with that agenda, PVAMU gave its first bachelor's degrees in engineering in 1945, covering architectural, civil, mechanical, and electrical engineering.[22] The nation's overall sense of its workforce needs and productive potential after World War II brought HBCUs an expanded expectation for their role in science and engineering. In 1947, mechanical engineering professor C. L. Wilson, the first African American to become a registered professional engineer in Texas, founded PVAMU's School of Engineering, of which he remained dean for the next eighteen years. In 1953, the college's directors moved the two-year trades and craft programs to a separate Division of Industrial Engineering, making a distinction vital to any attempts to establish a credible engineering school at the Prairie View campus. But PVAMU's ambitions for its engineering school may have outstripped its resources in this period. Due to funding shortages, a new building for the undergraduate engineering degree programs opened in 1950 without electrical and mechanical engineering laboratories that were part of original plans for the facility. Given the limited funding, the hiring capacities of the new engineering programs did not hold great promise: engineering faculty at PVAMU fell short of the level of preparation and experience seen at TAMU.[23] And, between 1959 and 1968, a greater proportion of PVAMU's bachelor's degrees were granted in industrial education (9.2 percent) than in engineering itself (7.2 percent).[24]

Ostensibly, the U.S. Supreme Court's 1950 decision in *Sweatt v. Painter* opened Texas graduate and professional programs to black citizens. However, neither the universities nor the court system pursued a thorough implementation of that change. PVAMU remained one of the few options for African Americans seeking higher degrees, and the school continued it efforts to grow in size and reputation. In collaboration with five other HBCUs, engineering faculty at PVAMU solicited the involvement of eighteen U.S. Atomic Energy contractors in technical research projects.[25] When NASA selected Houston for its home site in 1962, PVAMU, less than fifty miles from that city, began an enduring involvement with the space agency. But as a public university, PVAMU stressed its educational over its research function; it worked to obtain as large a portion of state educational funding as possible, reversing a history of severe neglect. In this regard the school's efforts to achieve equity through the civil rights era were guided by both practical and ideological concerns. Practically, PVAMU had long contended with a large state system of public higher education that was notably decentralized, leaving local interests to battle one another for shares of Texas' educational resources. Opportunities for smaller, poorer, and minority institutions in this mix improved, if only incrementally, as civil rights activism and then major federal legal reforms found a foothold in the state through Lyndon Johnson's increasing influence in Washington and Texas. Conservative Democratic governor John Connolly, very much a part of the Johnson political machine in Texas, created a body intended to streamline the sprawling higher education system: the Texas Higher Education Coordinating Board (THECB) was formed to distribute state educational funds in a newly efficient and "objective" manner.[26]

Despite profoundly integrationist sentiments on the part of Kenneth Ashworth, head of the THECB for over twenty years, Connolly's priorities in creating the board seem to have been a mixture of concern for disadvantaged populations in Texas and frustration with the ways that competing schools had grown independently without thought to how they might be duplicating offerings and thus wasting state money. Connolly recognized that the task of coordinating competing institutions would be "no easy burden" for the THECB, with individual schools likely to "cultivate, cajole, coddle and even brainwash" board members to act on their behalf. Part of his program for countering favoritism was the imposition of formulas for the distribution of funding.[27] With its share of the new public funding, PVAMU hoped to build up its provisions quickly in several engineering subdisciplines, but it made the "conscious and deliberate" choice to marshal its resources first toward the accreditation of its civil engineering program. The civil engineering department was the site from which all core courses in engineering at PVAMU were offered, but the focus on a single area, and thus the achievement of accreditation in at least one field, was intended by administrators also to provide a "psychological boost" for those working within the university. The university next developed electrical and mechanical engineering programs, obtaining accreditation for those in 1972. All three of these engineering units were considered by PVAMU faculty to be areas in which the university could rapidly build connections to regional professional networks and combat the economic marginalization long experienced by technical disciplines at the small minority-centered campus.[28]

The creation and staffing of engineering departments laid the institutional groundwork for the transition of PVAMU to a competitive school of engineering. At what level the school would actually function remained at that point unclear. There were several considerations to weigh. On a practical level, engineering programs are expensive to equip and operate; why duplicate existing programs at established majority institutions that had already begun to open their doors to black students in engineering? Through the 1950s and 1960s, as strict segregationist policies began to loosen, PVAMU engineering faculty often encouraged their students to transfer after two or three years to the state's white campuses, sometimes to College Station, but more often to the University of Texas at Austin, which had hired civil engineer Ervin S. Perry as its first African American engineering faculty member in 1964. Lamar University, in Beaumont, a city comprised of nearly 30 percent African Americans, eventually welcomed black students, as well, especially after 1970 through the vigorous recruiting and mentoring efforts of PVAMU alumnus and one-time faculty member, mathematician Richard Price. Both schools offered minority students immediate educational opportunities in reputable venues.[29]

Ideologically, even those who wished to enhance PVAMU's status as a site of technical work faced difficult decisions about the school's mission. Would the creation of relatively costly graduate and research programs, necessary for any engineering school hoping to achieve national stature, use public funding too narrowly, providing opportunities to fewer minority citizens than would other sorts of programming? Was the growth of educational opportunity for African Americans now so assured that PVAMU could turn away less qualified applicants? It is here that decisions made by the state regarding funding for PVAMU and the school's own sense of mission begin to intersect in exceedingly complex ways. As the state of Texas sporadically supplied new resources to the HBCU, PVAMU's leaders sought to make the best of that limited and unpredictable funding. Their vision of black inclusion in engineering is neither easily summarized nor judged in terms of its social radicality.

Ambitions "Versus" Resources

PVAMU began in the mid-1960s to be more aggressive in its pursuit of engineering students. When Thomas became president in 1966, he apparently gave his full support to Dean of Engineering Austin E. Greaux' inauguration of an "Engineering Institute." The institute brought high school students into contact with practicing engineers and instructors, closing what Greaux called the "believability gap" that had long dissuaded young African Americans from perceiving engineering as an attainable profession. Greaux clearly disliked the idea of distinguishing black students from majority cohorts in engineering, and decried "the stigma of 'the Negro Engineer'" and any programs that set "the Negro Engineer apart and separate from the total environment of the profession." Nonetheless, by the end of the 1960s, Greaux recognized that 90 percent of African American students training to become engineers were studying in "predominantly Negro institutions," and he urged members of the Texas Society of Professional Engineers to work as intellectual and financial advisors to those schools.

Many of his outreach and support initiatives for African American students grew and were centralized at PVAMU within the Engineering Concepts Institute (ECI) in 1971.[30]

The goals set by Greaux and others at PVAMU capitalized on a growing movement among U.S. businesses to bring considerable numbers of minority practitioners into technical positions, an impulse evident in the founding of the National Action Council on Minorities in Education (NACME) in 1974. PVAMU educators developed a wide variety of connections to build on that trend. Thomas saw professional contacts as linked to an enlarged research program and new graduate offerings at the university, with architecture and engineering as particularly important in the establishment of the school as a "portal" into "free enterprise." As the 1970s ended, PVAMU could turn to a variety of private sources for support for minority-led engineering projects, including programs at the Alfred P. Sloan Foundation which offered support to Prairie View, Howard, and four other "black engineering colleges" (as the foundation identified its grantees).[31]

Crucially, however, PVAMU was not likely to obtain considerable public funding for these efforts from its own state's coffers. The ECI program, for example, could not call on the state legislature with any expectation of support, but only on private foundations or corporations (in this case, General Electric) already committed to funding minority engineering programs. In 1980, when Thomas sought to establish an Office of Sponsored Research at PVAMU, a basic requirement for the operation of any competitive research institution, the Texas legislature offered him far less funding than he believed was needed, and that only on a year-by-year basis.[32] If we try to gauge the success of Thomas's programs in regard to either the employment prospects of students or the institutional growth and longevity of engineering at PVAMU, it appears that something in this recipe for inclusion was not working and that perhaps more funding alone might not have made any difference. Focusing only on the statistics that conventionally measure educational success, it is possible to conclude that Thomas was failing on his promise to "uphold standards" while pursuing inclusion. For example, freshmen entering PVAMU in 1983 had the lowest SAT scores in the state by a wide margin (the state average was 848; PVAMU's incoming students averaged scores of 660). What is more, in the early 1980s, PVAMU students showed a 75 percent failure rate on the Texas state engineering exam, compared to a 21 percent failure rate in the state overall.[33] But, as we have seen in the earlier historical cases, many of the structural conditions under which Thomas labored discouraged wide or complete implementation of reform strategies. Texas in these decades was not an environment in which publicly funded educational and professional opportunities for black citizens would be maximized.

By custom, if not by strict or consistent legal interpretation, Texas legislators through the 1970s denied to PVAMU the same levels of support (based on student numbers, among other factors) supplied to TAMU and to the University of Texas. This denial was maintained by a system that was as complex as it was arbitrary. Those predominantly white universities were maintained through the use of two state funds, the Permanent University Fund (PUF) and the Available University Fund (AUF). The AUF represents proceeds from interest and the sale of bonds held in the PUF. These resources were based on revenues derived from some two million acres of publicly owned oil and gas lands in Texas and other reliable sources of income. Historically, TAMU received one-third of available earnings from the PUF, while the University of Texas received two-thirds; AUF appropriations were variable but did not include PVAMU with regularity. Prior to major legal reforms finalized in 1983, PVAMU, despite its affiliation with TAMU, received no ongoing or significant share of earnings from any of these funds except for small amounts irregularly appropriated for it from PUF bond proceeds. PVAMU's eligibility for a set share of the AUF would have come from official designation of the university as the "college or branch university for the instruction of the colored youths," as required by the original Texas constitution. But as late as the 1980s, many politicians claimed that that designation had never been made official, and thus AUF funds need not be shared with the minority institution. Such relegation severely limited PVAMU's access to public educational funds. In 1980, for example, the University of Texas received $46.5 million from the AUF, while TAMU received $22.6 million; PVAMU received only around $3 million through indirect allocations, far less than a proportionate share on the basis of student population.[34]

The physical results from this lack of support were obvious at every turn. As the *Houston Post* wrote in several articles about the Prairie View campus in 1982 and 1983, a "facelift" was badly needed. Even successful growth initiatives on the 32-acre site, such as that which generated the $2.6 million engineering building in 1978, faced uncertain prospects: the completed building did not open for eight months because no money could be found for equipment or furniture. One visitor to the campus saw pervasive neglect and dilapidation, where even "most of the clocks and elevators" did not run.[35] State Representative Ron Wilson saw the campus as presenting conditions that were "worse than a community college."[36] In 1982, advocates suggested that PVAMU receive one-sixth of the usable money in the AUF fund, which would have amounted to about $22 million annually. Plans formulated around this time included the creation of ten new science and engineering programs. They projected the Prairie View campus in coming decades stripped of its dilapidated classroom buildings and outdated bull sheds and swine houses, and fitted instead with new facilities including laboratories for its architecture, engineering, and animal science programs; modern dormitories; equestrian and other recreational facilities; and even a continuing education center, motel, airstrip, and heliport.[37] But some analysts argued that PVAMU deserved no additional funding at all. They pointed out lax administrative practices at the school in this era; at least one case of poor cash management was documented by local media. Texas A&M System Chancellor Arthur Hansen saw no deliberate malfeasance but nonetheless judged PVAMU's administration over the previous decades as having "had a good heart" but having "gone downhill."[38] Significantly, both interpretations fail to acknowledge in any explicit way the difficulty of maintaining an institution with inadequate funds.

Two Texas journalists, looking back in 1983 at weak student and faculty performances at PVAMU over previous decades, proposed that "mediocrity was accepted by blacks and whites in education and the state Legislature."[39] But how are we to differentiate between a failure of vision and a realistic acknowledgement of limited resources? In 1982, Thomas argued before the state's Legislative Budget Board for a "dramatic financial effort" on behalf of PVAMU. He asked for $33 million dollars in appropriations, triple the amount allocated the previous year. As he tried to compensate for "80 to 90 years of neglect," it is difficult to see his case as weakly made.[40] Nor are the practical daily choices of PVAMU administrators during these years easily interpreted as negligent or misguided. One journalist, whose coverage of PVAMU was often negative, reported that Texas A&M System authorities had had to "force" PVAMU administrators to tear down a derelict building on that campus.[41] The A&M system vice chancellor for facilities planning and construction in this period similarly claimed that new landscaping had to be "forced on Prairie View officials." Both commentators imply a retrograde or lazy attitude among PVAMU leaders. But there is also evidence that those responsible for PVAMU's operations felt they could not count on receiving money when it was most needed, and their spending priorities may have been based on their experiences. When A&M system regents looked at the run-down campus and saw a failure on PVAMU's part to correctly allocate maintenance funds, they were also seeing the results of PVAMU administrators' decision to divert that money to student work programs for the 85 percent of students on financial aid, more than half from families living below the poverty level.[42] One PVAMU president confided to a state educational official that it was only by leaving two derelict buildings vacant and unrestored that he could hope to convince legislators of the severity of problems caused by underfunding.[43] It was not until the Texas state legislature, reacting to federal pressure and the insistent and creative activism of black leadership within Texas, altered the basic financial conditions of higher education in Texas that the campus would receive a considerable portion of the PUF and see true hopes of growth and any parity with College Station.

Federal Pressure and the Texas Response

As had been the case with many race-based educational issues through the 1960s and 1970s, progressive reform ultimately came through the efforts of the federal government. Some were simply direct efforts to offset generations of black occupational exclusion. For example, President Jimmy Carter initiated important new programs for HBCUs that encouraged federal agencies to establish small research grants and contracts which could be given to minority institutions without competition. The Science and Engineering Equal Opportunity Act, signed by President Carter in 1980, channeled federal funds to minority schools through the NSF.[44] Federal enforcement of civil rights legislation also played a considerable role in Texas' altered stance toward funding PVAMU. The federal government began enforcing provisions of Title VI of the 1964 Civil Rights Act (which prohibited discrimination in hiring, promotion, and admissions by any school receiving federal funding) across the country before the 1960s had ended. In 1969, the OCR notified ten states of their failure to comply with the Civil Rights Act. Texas received its own admonition in 1981. After a thorough investigation of the state's 100 institutions of higher learning, the U.S. Department of Education found that Texas' public universities had failed to "remedy the consequences of past discriminatory conduct."[45] State representatives at last grasped a powerful instrument for bringing about the redirection of state higher education funds to serve black Texans attending public institutions.

The threat of noncompliance rulings and costly litigation spurred Texas Attorney General Mark White and a number of white representatives to action. In early January 1981, the Texas Legislative Budget Board hastily recommended an "emergency" addition of $20 million to the 1982–1983 budgets of TSU and PVAMU (to receive $12 million and $8 million, respectively), while the University of Texas, the University of Houston and TAMU announced broad plans for increasing minority representation on their own campuses.[46] Near the end of 1982, after extensive efforts by

Representative Wilhelmina Delco and other voices of Texas' black communities, White was ready to endorse a constitutional amendment that would bring PVAMU into eligibility to share in the PUF, and he brought top administrators in the Texas A&M System along with him. Arthur Hansen, as the new A&M system chancellor, held an expansive vision for PVAMU that supported Delco's view. In a move with tremendous cultural implications, state leaders began to work on a new aspect of federal policy: the historically black university could expand and improve to attract more white students, and in so doing better serve all populations within Texas. At the end of 1982, the University of Texas and the Texas A&M System asked the State Legislature to authorize inclusion of Prairie View in the PUF.[47]

As they drafted the five-year Texas Equal Educational Opportunity Plan for Higher Education (generally referred to as the Texas Plan) for submission to the OCR in 1981, White and his collaborators envisioned a vast array of new facilities for TSU and PVAMU, including for the latter a 50 percent increase in usable space for the campus. TSU looked ahead to growth in law, health, accounting, and education fields. PVAMU, alongside anticipated growth in liberal arts, education, and health, refined plans for its ten new degree programs in science and engineering. New curricula proposed for PVAMU under the resulting final version of the Texas Plan and approved by the THECB included undergraduate degrees in computer science and in electrical and mechanical engineering technology. Those programs, not yet common in U.S. higher education, anticipated a rise in the use of computers, both as stand-alone technologies and as elements embedded in machinery, automobiles, and medical technologies. PVAMU also included new agricultural programs in its plans, but the new engineering technology programs were intended to replace the university's retrograde industrial arts programs with emerging "skills of the future." From this vantage point, modernization of the campus could accommodate traditional land-grant training and service agendas yet place a strong new emphasis on cutting-edge technological preparation, moving beyond the preparation of black students for careers only in their communities of origin.[48]

A shift in leadership for PVAMU occurred at this point. The Texas A&M System Board of Regents asked Thomas, who had been president of the university for 16 years, to resign from his position. He was reassigned as vice president for PVAMU's nursing programs, located in Houston. Ivory Nelson, a chemist who had served as Assistant Academic Dean and Vice President for Research at PVAMU, then became acting president for six months, during which time he oversaw the first stages of planning for a vast array of new facilities. The OCR-mandated upgrading of PVAMU was underway.

In January 1983, Nelson was replaced by a permanent appointment as president, Percy A. Pierre, said to be hand-picked by Hansen.[49] Pierre, trained as an electrical engineer, had previously served as Assistant Secretary of the Army for Research, Development, and Acquisition, and dean of the School of Engineering at Howard University. He had been a White House Fellow and an established consultant on urban educational issues, and he appears to have embodied Hansen's hopes for achieving a significantly higher overall caliber of student performances at PVAMU than that to which Thomas had apparently aspired. In Hansen's eyes, PVAMU, featuring a system of virtually open admissions for Texas students, had long represented a well-intended chance for poor black students simply to achieve exposure to an academic lifestyle, experiencing a kind of day-to-day formality and discipline they might otherwise miss. Historically, even as it developed engineering programs over the years, PVAMU had let occupational preparation rest at a subprofessional level for most students. This meant that if only a relatively small proportion of PVAMU graduates became fully qualified engineers, while most worked instead at the level of technicians, that was acceptable to the school's administrators. Characteristics of more prestigious HBCUs, such as Howard or Tuskegee, where Hansen had spent time, in his view were absent at PVAMU. Pierre, assured by Hansen that the resources needed to reshape the university would be made available, made explicit his own hopes to replace an agenda of "inclusion" with one of "aspiration," in which students attended college as an entry into the full range of professions.[50]

The University as Magnet

Pierre worked to extend improvements of PVAMU's physical plant, including the construction of a new $16 million library, but he also stated his intentions early on to use more than a million dollars in scholarship money to draw superb students from around the nation, much as the magnet model of public secondary education drew the highest caliber students from within a city to attend a single high school. He had helped implement this approach as an educational consultant and as a program officer for the Alfred P. Sloan Foundation in the mid-1970s, and singled out Houston's High School for the Engineering Professions, located within a predominantly black general high school, for particularly high praise as he planned PVAMU's future. In Pierre's recollection, this "school within a school" promised PVAMU "instant academic credibility," drawing strong students as well as faculty who might not have been attracted to a school with a less conspicuous agenda of high achievement. With better students and faculty PVAMU would be better positioned to attract significant research funding as well.[51]

A primary goal for Pierre was the narrowing of PVAMU's "brand" from a broad but shallow array of offerings (before which, as he put it, "presumably, Prairie View could prepare one for everything from the research laboratory to the beauty salon").[52] The establishment of the Benjamin Banneker Honors College (BBHC) at PVAMU captures Pierre's vision most aptly. Upon his appointment, Pierre proposed the creation of a special residential unit within the university for highly motivated students who would live together in a single dormitory and take part in an accelerated, academically challenging curriculum. SAT scores and class ranking would help the college select high-achieving students likely to persist in the enhanced curricula. The main objective of this honors college, named for the eighteenth-century African American mathematician and astronomer, was, according to Pierre, to have students "graduate with the credentials that will get them into the most selective graduate and postgraduate programs." Science and engineering were to form the central elements of the new college, at least in its early years. In the mid-1980s, a minute proportion of PhDs in science, engineering, and mathematics were granted to African Americans: only six such doctorates were awarded in mathematics in 1986, and eight in physics, compared to 462 in education disciplines that year. Expanding the pool of minority graduate students would have direct effects on the hiring of minority faculty in the sciences and engineering, which in turn might draw even more minority students into these fields. In anticipating still further ripple effects, Pierre envisioned the honors college as inciting enhanced programming and a generally heightened level of academic discourse and activity across the campus.[53]

Fourteen other honors colleges existed in U.S. universities at this time, but this was to be the first at an HBCU. The Texas Higher Education Coordinating Board approved creation of the honors college at Prairie View in 1984, and the college established a core curriculum—honors versions of many other undergraduate courses—and a dormitory exclusively for BBHC students. Planners felt that the residential setting would "complement the instructional programs" and assure a distinct sense of identity for the exceptional students. The college's first dean was Ronald Sheehy, an administrator and microbiologist from Morehouse College, under whom BBHC opened offering two majors: biology and electrical engineering. Operating from 1990 to 1998 under Dean Jewell Prestage, a political scientist who studied patterns of minority occupational participation, the honors college firmly established higher standards for student admissions than the university previously had held, a major national recruitment program backed by scholarships, summer and co-op programs, and collaborative projects with Big Ten and Ivy League universities in a number of disciplines.[54] Within a few years, the college had established a milieu in which students in the hard sciences and engineering worked in many of their classes beside students in education and the liberal arts, with shared exposure to many of the philosophical underpinnings of classical university curricula.[55] Prestage had been active in honors education before her appointment as BBHC dean and made a priority of linking PVAMU's honors program to others through the National Association of African-American Honors Colleges and regional organizations. Both Prestage and Pierre hoped to equip minority students for an unrestricted range of occupational, civic, and economic opportunities. Both also likely understood that honors colleges might have a "halo effect" whereby the hosting institution gained legitimacy and resources for all of its programs.[56]

In several respects BBHC's planners had an extremely strategic sense of the college's mission. Its focus on the labor market led the honors college to offer its first degrees in electrical but not civil engineering as once planned, when employment conditions showed civil engineering to present lower demand than the other. The university at this point maintained one college for Applied Science and Engineering Technology and another for Engineering and Architecture, reflecting a conventional stratification of technical work into two levels of professional preparation. But the new honors college, geared toward the preparation of students for graduate work, now could solidly identify PVAMU with the most esoteric spheres of science and engineering practice. The bachelor's degree had long represented the professional degree in engineering, and in deciding to cultivate graduate candidates in engineering disciplines, PVAMU bypassed the obvious developmental step of simply improving its undergraduate programs. Instead, it aimed for the most elite reaches of engineering education.

As Ivy League universities in this period presented a source of "fierce competition" for the nation's best black applicants, successful honors colleges in minority institutions addressed the challenges of both educational equity and institutional survival. BBHC could attract excellent students to Prairie View from around the country. The honors college soon found itself drawing large numbers of students from certain regions, such as the Detroit area, as word of mouth about the college reached predominantly black high schools. Students who might otherwise have chosen to go to majority institutions could see in PVAMU an exemplary set of programs offered in a historically black setting; between 1983 and 1987, the university's overall SAT scores among incoming freshman rose by 120 points. In some years of its operation, the proportion of BBHC graduates who enrolled in graduate or professional programs approached 100 percent (albeit many in non-science or engineering fields), and it would appear that the college had found its niche.[57] The association of a black-identified institution with the highest levels of educational achievement represented for Pierre a distinct path toward correcting educational inequities of the past. From a few quarters came the claim that racial bias could best be overcome by eliminating the whole category of "black" universities, thus relegating HBCUs to an outdated relic of segregation. But that proposal was often put forth disingenuously by some extreme conservatives who presumed that black Texans would not move to white universities, choosing instead to drop out of public higher education altogether if black institutions ceased to exist. While such schemes garnered attention from the press, Pierre recalls that almost no one in Texas politics or education in this era took such suggestions seriously.[58]

Instead, Pierre found a receptive audience for his idea of crafting a competitive academic enterprise that was also a place of heightened black identity. He believed that the cultivation of racial consciousness among black students could be conducive to learning, and he worried when black enrollment at PVAMU dropped in the mid-1980s, even as the campus was fulfilling its mission as outlined in the Texas Plan to attract more white students. This plan required TSU and PVAMU to increase nonminority enrollments, an effort which resulted at PVAMU in newly liberal admissions and transfer policies and in special recruitment efforts aimed at majority audiences. The number of white students on the campus increased from 271 to 426 between 1981 and 1985.[59] Computer science, an appealing field for those seeking employment in the growing high-tech industries of nearby Houston, showed a particularly pronounced increase in the number of white students, possibly because of the ease with which Houston residents could commute to Prairie View.[60] But Pierre did not want any lessening in the campus' sense of its own black heritage or in the feeling of community that he and other analysts believed a perception of shared heritage would engender. Like an increasing number of analysts watching the diminution of civil rights activism in 1980s America, Pierre understood the challenges that majority institutions posed to minority students, and he saw the HBCUs as offering many advantages in terms of students' social and emotional well-being.[61]

The Engineer, the Individual, and the Collective

As Pierre, Prestage, and other energetic faculty and administrators at PVAMU built up the honors college, the whole question of where race, and indeed, social identity, should figure in higher education was in flux across the United States. This was nowhere more so than in science and engineering, a set of academic disciplines that historically minimized open address of its practitioners' social experiences. For some educators who recalled campus uprisings around the country in the 1960s and 1970s, student activism of any kind interfered with sober and industrious academic work. In many ways even the least extreme strands of the Black Nationalist and other separatist movements posed particular threats to science and engineering fields. On a philosophical level, those ideologies might challenge the basic precepts of industrial capitalism, and thus taint the prospect of careers in industry, military, or government spheres for engineering students. But some educators seemed to have detected a more basic incompatibility between pronounced black identification and occupational achievement. Thomas himself seems to have equated heightened black consciousness with a potential for socially disruptive impulses. Summarizing his concerns in 1977, he reflected, "We've created a lot of black pride, which should've been done a long time ago, but we also created a lot of things that weren't too helpful because we created the impression that just being black would do it."[62] As Thomas saw it, Prairie View graduates, upon achieving economic security, ideally would become "integrated into the American stream as middle class Americans." This meant that, "as more blacks join the so-called middle class, fewer blacks become prison inmates and stand in welfare lines ... fewer blacks express anger as they did in the 1960s."[63] Surely Thomas understood the role that black institutions played in the entry of minority practitioners into the professions: in 1970, three-fourths of all blacks who held doctorates had attended HBCUs. But he also seems to have favored a systematic de-emphasis of the uniqueness of black experiences in the educational setting, as if black political radicality somehow equated with economic insecurity. For Thomas, success would come not through federally imposed "quotas" that might bring more minority students into higher education, but through each student choosing to be "the quintessential Horatio Alger hero and overcome all obstacles—financial, spiritual, physical."[64]

That vision of personal development and uplift through higher education was compatible with Thomas's inclusive ideology at PVAMU; the university would provide a means through which even weaker students might develop their inner resources and thus find a route toward academic and professional attainment. But it is not an outlook easily characterized as politically transformative, falling back as it does on individuals' inner resources. Thomas's critiques of race activism map readily onto the historical characterization of rigorous intellectual work as essentially an individual pursuit, with which Pierre and other advocates of black-identified higher education had to contend.

Employers of technical personnel also customarily support the primacy of individual progress. In each of these ways—through conventions of grading, reward systems, and professional rhetoric—the field of engineering discouraged the connection of students to collectivist social movements through the last decades of the twentieth century just as it had in the context of race activism in the 1960s. But for Percy Pierre and other supporters of PVAMU's honors programs in science and technology disciplines, race consciousness for black engineering students did not seem antithetical to the highest levels of technical study and research. Complicated questions about the reformist potential of engineering begin to come into focus when we examine their actions. Like educational analysts Patricia Gurin and Edgar Epps, in their 1975 survey of black "consciousness, identity and achievement," Pierre saw no conflict between "relevant identity goals and the traditional academic goals of higher education." Gurin and Epps found that while "normative pressures in college generally push toward individual development, expression and advancement," under certain educational conditions individual and collective goals might be reconciled. Black self-identity among college students—whether expressed through militant political action, demands for curricular inclusion of black history and culture, or a personal focus on ethnic heritage—might lead to a *stronger* commitment to one's individual life goals: "Collective achievements promote the collective aspect of identity. To identify with a group whose achievements you admire and, furthermore, to enhance those achievements by your actions cements the mutual relationship between the individual and the group, the personal and collective elements of identity."[65] Gurin and Epps cited many cases in which minority students both gained "the skills necessary to act effectively in the world" and "developed the sense of identity and humanity that provide the reason to act." For Pierre and others, BBHC seems to have promised such a combination.[66]

Importantly, unlike some educators' invocations of black leadership in earlier decades, which implied separate economic and civic spheres for white and minority Americans, Pierre's vision for PVAMU's science and engineering programs imposed racial identity on a unified realm of professional practice. Thus, the HBCU could perform a function in diversifying technical occupations rather different from that of predominantly white universities. Percy Pierre envisioned a combination of academic excellence and racial identification, but his conception of the former followed some traditional definitions: BBHC confined admissions to students arriving at college with "a strong college preparatory background in math, science and English."[67] In other words, as in Thomas Martin's presidency of IIT a decade earlier, the honors college at PVAMU projected exceptional new technical opportunities for "qualified" but not "qualifiable" minority students. That sort of selectivity, as we have seen, might perform exclusionary work and reinforce existing differentials in opportunity for historically advantaged and disadvantaged aspirants. Because it leaves eligibility criteria unquestioned, it may also leave discriminatory social-structural conditions in place. However, selectivity of this kind functions very differently when it is closely associated with minority identity. Pierre's educational vision had tremendous positive implications for larger issues of black identity, interleaving elite achievement with social agendas centered on collective sensibilities. In so doing, BBHC linked intellectual and social projects in ways projected by the most ambitious reformers of the 1960s and 1970s.

After "Banneker"

By the mid-1990s, the Benjamin Banneker Honors College had an enrollment of 408 students drawn from thirty different majors, and in 1995 it graduated thirty-one students with full honors credentials. It is not clear from university records why more students were not completing the entire honors curriculum; having operated for nearly ten years, it would seem that something closer to a quarter of its enrollment should have been graduating each year. Of those who did graduate, eleven went on to graduate or professional school.[68] Given the original mission of the honors college, this is a small number in absolute terms considering that BBHC was by now entering its second decade. It is not clear why this was so. Certainly the faculty recall a very high caliber of work being done by honors students, and innovative coursework arising from the college's interdisciplinary design.[69] Many BBHC alumni eventually moved into high-level corporate and government positions, and surely many of these people showed great promise even in their first years out of college. Yet PVAMU leadership may have been gradually moving away from whole-hearted support of this selective program, and institutional obstacles begin to appear in this period that may have undermined the operations of the honors college.

Pierre left PVAMU in 1989 and was replaced by Julius Becton, a former lieutenant general in the U.S. Army with a background in economics and experience as director of the Federal Emergency Management Agency. BBHC appears to have had ample institutional support for most of Becton's administration, but around the time of his retirement in 1994, a different pattern emerges. Enrollments in the honors college were still growing, but BBHC's annual report of 1994 indicates a series of problems ranging from inadequate funding to emerging logistical challenges. Because university scholarship money, the lifeblood of BBHC's recruiting efforts, was often delayed in reaching incoming students, many chose to attend other universities after accepting offers from BBHC. The *Annual Report* also noted a significant administrative development encountered by the honors college: its operations were moved in the spring of 1994 to smaller quarters, identified in the *Annual Report* as "the least desirable office space in the entire academic area at the University." Despite a campaign to raise money for the construction of a dedicated building for the college and serious plans for growth voiced by its advocates, BBHC seems to have been facing diminished prospects.[70]

When Charles Hines, a retired major general in the U.S. Army, was appointed as PVAMU president in the fall of 1994, and for two or three years thereafter, the honors college felt some sense of renewed support, but it was under Hines's leadership that the university ultimately dismantled BBHC in 1998. The faculty of that era make reference to only very general reasons for the closure, such as a "change of emphasis" by PVAMU leadership, or "presidential prerogative." A report prepared by the Texas A&M System for the OCR in 2002 dismisses the closure of BBHC as simply a matter of "changing times."[71] But one faculty member who taught in BBHC in the mid-1990s and followed its progress thereafter believes that Hines may have had an alternative vision for PVAMU into which an honors college may not have easily fit. That vision may have been focused more on immediate employment prospects in the service sector than on a wider range of post-graduate or professional opportunities for PVAMU degree holders. What is more, state legislators encouraged the development of new programs at PVAMU only if existing expenditures were reduced. Closing the honors college allowed Hines to save the cost of a dean's salary and the college's other administrative costs, and also to use existing PVAMU faculty to teach larger, non-honors sections.[72]

It is difficult, however, to construe the closure of BBHC as merely an economic or practical decision for the university. Hines may have had his own thoughts about optimal student audiences for PVAMU; audiences that would not necessarily be drawn by the presence of an honors college.

Certainly it would have been obvious by 1998, two years after *Hopwood* drastically undercut racial preferences in Texas university admissions and set-asides, that the state was losing exceptional minority students to out-of-state institutions which had begun aggressively recruiting in Texas. PVAMU's honors program might have countered some of those trends.[73] In several instances, Hines seems to have targeted the highest level of technical pursuits at PVAMU for cuts or cancellations. Hines carried a reputation of having a distinctly top-down management style, notable especially in conflicts with faculty and staff who believed that he put little importance on the traditional university ideal of shared governance.[74] Hines's failure to accommodate faculty concerns became one more data point for analysts who saw a pattern afflicting HBCU governance, whereby policy-making boards chose school presidents inclined to serve their own interests, making such appointments without sufficient faculty input.[75]

I will return in Chapter 7 to the conditions under which engineering research proceeded at PVAMU. For the moment, it is not necessary to ascribe ill-will to Hines to see that closure of the BBHC could erode the newly selective reputation of PVAMU and run counter to the interests of faculty who had worked hard over the previous decade to build up its programs. Whatever his motives, closing the honors college complied with a prevailing political distaste in the late 1990s for the expansion of pedagogically ambitious racially identified programs. In 1996, the TAMU system readily complied with the passage of the *Hopwood* decision by terminating race-based scholarships and admission decisions. In a legal and social climate turning against race-conscious public policies, PVAMU leadership expanded its "University Scholarships," intended to recruit and retain excellent students, as a replacement for the honors college. That individualized program, which was meant to attract high-achieving students and integrate them into the mainstream curriculum, was for some reason preferable to the maintenance of a free-standing educational unit or special curricula designed to serve exceptional students.[76] At PVAMU, this shift eliminated one conspicuous connection between academic excellence and black demographics. Comments of some PVAMU faculty at the time suggest that Hines saw the HBCU's role as one of more traditional inclusivity, returning to the pre-Pierre model of serving Texas youth at that traditional "point of greatest need."[77] Certainly, greater inclusion at all levels of academic achievement was needed. In 1999, black Texans still fell behind the state's high school graduates overall in bachelor's degrees.[78] Seen in those terms, the maintenance of an honors college connoted perhaps too small an intervention. Yet at least one alumnae, who later became a staff member at PVAMU, in 2005 associated the closing of BBHC with a diminished "family feeling" at the university. He expressed hope that the honors college might be reopened as a way of recapturing a sense of black identity that the campus had lost as more nonminority students had enrolled. There is, in fact, some indication that such a college at PVAMU is now under consideration. But if Hines was in fact hoping to redirect PVAMU's mission away from a heightened profile in the world of black-identified scientific achievement, closing the honors colleges was one way to accomplish that goal.[79]

Conclusions

When members of the PVAMU community worried about the loss of black identity through shifts to a more diverse student body and the elimination of high profile, black-identified academic programs like BBHC, they reflected on a major current in U.S. race relations. In its effort to address imbalances in minority and majority public education through the 1970s, the federal government pushed systematically for the introduction of white students into traditionally minority universities. If Texas was to avoid duplicating programs among its publicly supported universities, and thus truly dismantle the last vestiges of the segregated dual system, PVAMU and Texas Southern University would need to bring in white students. In 2000, the OCR, assessing Texas' progress toward integration, expressed concern that without more unique programs the predominantly minority schools offered insufficient attraction to majority students.[80] That priority in federal education policies, to integrate not only historically white but also historically black, Hispanic, or Native American campuses, suggests that diversity meant not necessarily the enhancement of minority self-identity for students but possibly its diminution. If the political implications of maintaining largely black schools were too radical for some tastes, voices on both the left and right had difficulty understanding the "puzzling" requirement of Title VI to enhance educational provisions at predominantly black institutions while desegregating higher education. Some analysts had seen these "conflicting goals" as generating federal HBCU policies that were "vaguely proposed and variously enforced."[81] Seen in the context of a single state school system, those two requirements reveal profound ambiguities in official U.S. ideology about ending race-based educational disadvantage. On one level, states like Texas that had maintained markedly uneven institutions for majority and minority students had to find the means to enact change in both kinds of settings, raising the problem of resources discussed at the beginning of this chapter. But resources aside, it is difficult to see the likelihood of profound change in what was unquestionably a time of ascendant conservative governance. Fewer federal interventions displayed the clear purpose of redressing historical race-based inequities.

As was true of the previous case examples, here one finds interlocking ideologies about technological modernization and social stasis. In 1979, Lorenzo Morris summarized the nation's failure to supply a secure source of funding to traditionally black institutions. He made the important and rarely heard point that in stressing scientific research and development as the main criteria for federal funding of universities, the country had put most of the HBCUs out of play for such funding and thus denied support to "those few institutions that had fundamentally committed themselves" to equal opportunity.[82] What is more, federal legislation providing direct aid to students remained silent on connections between disadvantage and race, limiting the possibility that historic discriminatory patterns in majority college admissions might be recognized. When the OCR found traces of segregation after studying PVAMU and TSU in 1999, then-Governor George W. Bush worked with the OCR and the THECB to reach an agreement about how to remedy the situation. The resulting state-appointed panel created "Priority Plan 2000," laying out recommendations for new scholarships, fundraising, and increased enrollment as well as increased emphasis on technology applications across the curricula of both schools. Doubts about the efficacy of these latest plans came from many corners. Some saw Bush's concern for the HBCUs as an expedient for his upcoming run for president. Democratic lawmakers from Houston, and former-PVAMU president Julius Becton, all believed that the new plan, the fourth in twenty years, was "modest," and that it merely echoed older ineffectual attempts at achieving racial equity in Texas higher education.[83] Not everyone found the plans to be lacking, however. Republican supporters of George W. Bush, as he prepared to run for president, accused OCR head Norma Cantu, studying what she saw as PVAMU's low status compared to TAMU's for the OCR's 2000 finding on the state, of conducting a "hatchet job" meant to embarrass the aspiring candidate.[84] But in 2003, as PVAMU prepared to begin operations under Hines's successor, historian George Wright, civic leaders and journalists noted that Wright would face the "challenge of reviving" the school, still far from the status of Florida A&M or North Carolina A&T universities.[85] Decades of effort had not yet corrected generations of racial discrimination in the Texas A&M System.

With the roller-coaster history of competing ideologies at PVAMU—sometimes striving for a broad-based kind of inclusion, at other times pushing for exemplary academic achievements—it becomes clear that no single vision for the black land-grant school prevailed even after civil rights legislation. Under each administration at the school since the 1970s, and in a legal climate increasingly averse to affirmative action and other race-based intervention, there is no question that ideas about professional eligibility in engineering involved complex recipes for achieving competitive stature. One may understand the complexity of scientific and social interconnections in contemporary higher education by looking closely at two other efforts of the Texas A&M System to address racial diversity, both of which allowed for flexible definitions of competence that in turn brought membership to groups previously excluded from engineering realms. Minority Engineering Programs (MEPs) on the College Station campus after 1980 displayed innovative teaching and learning schemes that pushed conventional ideas of merit to allow, if not greater inclusion in absolute numbers, at least the admission of students of more varied backgrounds and strengths. And, even as Hines hesitated about the role of high-tech research at Prairie View through the early 1990s, groundwork was being laid in several PVAMU engineering programs for truly competitive projects to be conducted under federal auspices. In examining the institutional and epistemological conditions surrounding those projects, we may learn more about the susceptibility, and resistance, of engineering to social reform.

CHAPTER SEVEN

Standards and the "Problem" of Affirmative Action

Departures from Convention in the TAMU System

THE IDEA THAT RACE-BLIND POLICIES held the key to making education and employment in the United States more nearly fair, more essentially *American*, steadily gained credence after 1990, shaping even the rhetoric of some relatively liberal interest groups. Analysts who persisted in identifying racism as a source of economic inequity often faced critics who claimed that this persistence was part of the problem. To such critics it was only the erasure of race, in public discourse and thereby, presumably, in private ideology, that would move the nation forward in the post–civil rights era. In this atmosphere, bent on masking the complexity and structural foundations of U.S. race relations, the detection and measurement of individual merit gained new centrality in ideologies of U.S. higher education.[1]

The recognition or celebration of ethnic heritage did have a place in a newly refashioned agenda of diversity that ostensibly lent value to cultural difference, but only when the general tenets of a competitive meritocracy were in place, with every student to be judged and rewarded solely on the basis of his or her performance. These strategies denied that race still functioned as a determinant of students' life experiences, likely to bring privilege to some and disadvantage to others. Even judicial decisions that upheld the selective use of race in university admissions found an unsure purchase in the academy after 2000. Through its selective promotion of only those legal findings that were critical of affirmative action policies,

George W. Bush's presidential administration discouraged university efforts geared expressly toward support of minorities. By 2005, many universities had moved away from recruitment and retention plans based on students' gender, race, or ethnicity in the often inaccurate belief that to do otherwise was to court legal action. In all but the most contained expressions, those categories of personal identity were newly stigmatized as objects for both educators' and students' attention.[2]

In the public universities constituting the Texas A&M System, that stigma held particular power. In a partial reversal of some of the most stringent anti-affirmative action rulings of the previous decade, the U.S. Supreme Court ruled in 2003 that under certain conditions universities were permitted to consider race in admissions and distributions of financial aid. With these so-called Michigan decisions (which arose from suits brought against the University of Michigan's law school by white students who had been denied admission), *Hopwood,* which since 1996 had undermined efforts at racial diversification in Texas, lost much of its legal power. Within hours of the 2003 decisions, the University of Texas began to reinstate admissions and other programs aimed at women and minority students. But Robert M. Gates, then president of Texas A&M University at College Station (TAMU), thought better of those options. Instead, he promised aggressive and targeted recruiting that would bring minorities to TAMU without recourse to the adjusted academic standards implied by race-conscious policies. Given enough time, money, and personnel, he believed, he could establish outreach programs that would draw an enlarged pool of highly qualified black and Hispanic young people to College Station. By the time he left TAMU to become Secretary of Defense under Bush three years later, he had enacted his plan: unprecedented numbers of minority students enrolled in TAMU in 2005.[3] But critics point out that this increase seems sizable only in relative terms, because the campus had been neglectful of minority recruitment for generations. In 2006, African Americans made up only 3 percent of undergraduates at College Station; Hispanics roughly 10 percent.[4] *Eligibility* for participation in higher education at College Station in science and engineering fields, which comprised many of TAMU's signature programs, did not expand. Black and Hispanic Texans from weaker urban and rural high schools, who among Texas high schoolers were the least likely to have had optimal preparation in math and science areas, would find little place in the newly "diversified" TAMU. No matter how aggressive recruiting tactics might become, eventually the number of minorities entering TAMU would plateau.[5]

While president of TAMU, Gates stated frequently that diversity was one of his priorities for a campus long known for demographic homogeneity.

The university appointed its first vice president for diversity under Gates. Yet, in denying the necessity for legal interventions into social relations, Gates confined his efforts to those measures that sustained conventional ideas of competence. As he put it: "students at Texas A&M should be admitted as individuals, on personal merit—and no other basis." According to Gates, family hardships and other sources of personal adversity would be given new weight in TAMU admissions decisions, evoking PVAMU President A. I. Thomas's reference to Horatio Alger some thirty years earlier. But given the option of using race in admissions, Gates chose not to, implying that his institution as well as its students would fare better without recourse to that category.[6] In this formulation of student identity, a sense of oneself or others as members of certain social collectives (such as that embodied in the traditional "Aggie" spirit that united many TAMU alumni) was to be cultivated, while other shared self-understandings (such as racial or gender identity) were not. Alongside avowals of incompatibility of ethnic self-identification and academic achievement among disadvantaged minority students, the predominance of white and Asian or South Asian minorities in technical occupations was to go unremarked.[7]

In suppressing attention to race, this "modernized" ideology buttresses the discriminatory conditions still faced by many black Americans in various professions today. One need not see Gates's subsequent efforts to "persuade more minorities" to enroll at TAMU as having been offered in bad faith or naivete. Rather, Gates simply denied the significance of some forces responsible for the historical absence of black students at College Station. His words echo the vulnerability ascribed to science and engineering in the 1970s when universities reacted to diversification pressures with warnings about "lowered standards." As seen in the earlier case studies, curricular standardization itself has a powerful social instrumentality.[8] In primary and secondary education after 2000, Bush's "No Child Left Behind" programs carried this ideology into communities across the nation, while at the college level, anti-affirmative action arguments heard since the 1960s about maintaining admissions and classrooms standards showed renewed vigor. To comprehend the likelihood of change under the new Obama administration, it will help to understand that while some university administrators and political leadership foreclosed the discussion of standards as a cause of economic inequities through the 1990s, other educators resisted this trend. Within the Texas A&M System, at College Station and Prairie View, engineering faculty, staff, and some supportive administrators have developed teaching and research programs that restructure prevailing practices believed by educators to contribute to raced and gendered patterns of occupational attainment. Their efforts often are supported by federal initiatives which remain, even in an era of a greatly diminished role for government, a source of progressive change in U.S. higher education. Two such projects that seek to channel federal funding into new opportunities for minority engineers are considered here. In the first, the National Science Foundation (NSF) has provided money to innovative teaching projects in technical fields at TAMU. The second has earmarked portions of federal research-agency funding for HBCUs—in this case the provision of NASA funding for engineering and science projects at PVAMU.

Both initiatives are based on flexible conceptions about the structure of technical work. In the first portion of this chapter, I describe curricular reforms undertaken by TAMU's Dwight Look College of Engineering after 1990 as part of the NSF's nationwide focus on engineering education. Racial, ethnic, and gender diversification have constituted a central goal of these NSF efforts; I will probe to what degree and with what results TAMU's programs embodied that agenda. As a member of the NSF-funded Louis Stokes Alliance for Minority Participation (LSAMP), and particularly through its involvement with the NSF's Engineering Education Coalitions (EEC) program, the college through the 1990s embraced relatively flexible notions of what constituted meritorious technical work. Innovative faculty and administrators at College Station achieved nationwide influence as they implemented new ways of teaching undergraduate engineering courses, often addressing issues of individual and collective experience in ways that potentially enlarged the role of racial identity in engineering pedagogy.

Returning to Prairie View in the second portion of this chapter, I will examine a similarly innovative set of ideas about technical competence and inclusion, in this case regarding not engineering pedagogy, but eligibility for federal research funding in science and engineering. Beginning in the 1980s, major federal agencies involved with defense, health, and technical research designated portions of their budgets for use by the nation's minority-serving institutions, bringing unprecedented support to science and engineering work at those schools. NASA's work with HBCUs departed from conventional research-funding procedures, resulting in important scientific contributions by these historically underserved institutions. In PVAMU's engineering research efforts developed under NASA's auspices, constructions of competency and competitiveness (among students, faculty, and researchers) have been realigned, and minority practitioners have been welcomed into scientific spheres from which they were previously absent. As in TAMU's engineering curriculum reforms, a picture emerges of a world of practice in which openness regarding definitions of rigor is cognate with social openness.

Diversity at College Station: "Welcome to Aggieland"

The Texas Plan approved by the Office of Civil Rights in 1983, discussed in Chapter 6, called on the Texas A&M System to pursue what seemed to many to be conflicting goals: enhancing a historically underserved black-identified institution at Prairie View, while at the same time drawing exemplary minority students to its white campus at College Station. Kenneth Ashworth, head of the Texas Higher Education Coordinating Board in this era, saw those goals as opposed but nonetheless equal priorities for the state and its public higher education system. "We just did it," he recalls, describing the board's work as a matter of taking all of its charges at face value in order to enact as many reforms that would serve as many students as possible.[9] Thus, while PVAMU aspired after 1981 to a considerably enhanced stature in engineering teaching and high-tech research with its new resources, and contended with the social meaning of HBCUs in America's changing racial climate, Texas A&M's main campus was pursuing the goal of ethnic integration. Set in 5,200 acres on the edge of Bryan, Texas, the campus at College Station served some 30,000 students and employed just under 2,000 faculty by the end of the 1970s. This student body was overwhelmingly white, with just 111 African American students (0.37 percent) and 244 Hispanic students (.83 percent).[10] In Washington, the OCR, unsatisfied with Texas' fulfillment of Title VI provisions, saw those numbers as falling far short of their goals for the state and pushed the Texas A&M System to diversify.

The Texas Plan outlined many concrete steps toward racial equity to be taken by the schools of the Texas A&M System, from the adoption of a newly inclusive philosophy regarding admissions and hiring to improved dormitories and classroom buildings. Few in Texas disputed that the land-grant schools were vital sites for such racial reforms in higher education and employment. Because almost a third of all TAMU students enrolled in the College of Engineering, changes in that college might be expected to have considerable effect on the instructional culture of the College Station campus overall.[11] But in what proportion would funds go to programs at PVAMU to build that historically black institution, or to TAMU, badly in need of new programs and services to assure its racial integration? Other campuses of the Texas A&M System and across the array of public colleges and universities in Texas, many of which were serving growing minority populations in the state, also exerted claims on educational funds, further complicating matters. There is some evidence that race itself did not take center stage for those debating this issue. Percy Pierre, president of PVAMU during the 1980s, believes that few whites outside of educational policy circles would have cared deeply about the HBCU's existence from anything other than the standpoint of how state resources might be distributed. Certainly, concerns about duplication of academic programs among public institutions have been a perpetual feature of Texas politics, as they are in virtually every state. But implicit in the competition among educational programs aimed at minority inclusion in this period is the fact that Texas chose not to maximize diversification efforts in its public universities. The level of the state's commitment to race-based educational inclusion efforts becomes particularly clear in and beyond the *Hopwood* years, between 1996 when that prohibition on race-based policies was passed and the 2003 Supreme Court decisions that partially repealed the ruling. Even after the Michigan cases permitted some consideration of race in higher education, sincere and innovative attempts to correct historic inequities at the largely white College Station campus played out in a general climate of ambivalence about social change.[12]

The two campuses could be seen as standing at extreme ends of an ideological spectrum regarding race. An HBCU might reasonably be expected to encourage racial identification among its students, as PVAMU had through the 1980s. Even as it complied with OCR mandates for increased white enrollment, PVAMU emphasized its ethnic legacy and unique character in the Texas A&M System. A campus striving to end old habits of racial homogeneity, as TAMU sought to achieve in this era, might shy away from such emphases. In fact, inclusive rhetoric at majority white institutions commonly outlined ostensible advantages to minimizing the "marking" of students by ethnicity or gender.[13] Furthermore, TAMU had long had a reputation of recurring racial tension and immensely powerful cultural traditions that to some observers held the potential to reassert the institution's exclusive social habits. Race-based programming might have revived historic divisions among students of different ethnic backgrounds. Minority-identified campus activities continued to incite white backlash even through the 1990s. By 2000, many boosters, including Gates, were openly claiming that the "Aggie" culture offered a shared experience that could "transcend diversity" and promote the espoused "race blindness."[14] How could Texas educators and lawmakers overseeing federal reforms at PVAMU and TAMU reconcile these divergent attitudes about students' self-identity? The ambitious curricular changes in TAMU's engineering programs, representing the NSF's socially informed goals for engineering education, test the depth and impact of these reformist projects.

Minority Engineering Programs at TAMU

With PVAMU's inception predicated on the "whites only" admissions policy at College Station, it is perhaps not surprising that racial integration at TAMU did not begin until 1963, when three African American students were permitted to enroll in summer programs there. Hispanic students historically had been admitted to the campus, but in low numbers. The Texas A&M Board of Regents may have admitted the African American students to TAMU in light of threatened court action at other segregated branches of the system to which black students had sought admission, but there are some indications of a general loosening of social strictures on the campus in this period.[15] Much of this occurred under the presidency of Earl Rudder, considered to be a great modernizer of TAMU, between 1959 and 1970: the wives and daughters of faculty members became eligible for admission to TAMU in 1964, with general admission for women beginning in 1970. In 1965, Rudder ended compulsory participation in the school's Corps of Cadets, a military training program which had lent to a pronounced culture of white, male camaraderie on the campus. While these changes may have seemed cataclysmic to those invested in longstanding customs on the College Station campus, TAMU remained a rearguard participant in racial diversification. Retrograde sentiments erupted in a number of events, including the burning of a cross at the dormitory of the first black student to start on the football team. As a land-grant school TAMU had long heralded its character as a place where "anyone can succeed," but as historian John Thelin has written, localized cultures can strongly influence the conduct of universities. In the case of TAMU, those cultures were inflected with racialist ideologies. TAMU did not hire its first affirmative action officer until the early 1970s, and enrolled only a tiny proportion of African Americans through the peak years of U.S. civil rights activism.[16]

For many at TAMU, open expressions of black consciousness were simply irreconcilable with educational mandates. Facing demands in 1973 from the student-run Black Awareness Committee, university president Jack K. Williams sought to dissuade the committee from pursuing the creation of a Black Studies curriculum. He urged that students instead go into engineering or other areas in which, he believed, employment would be forthcoming. Well intentioned as such advice may have been, it devalued the project of identity-based scholarship and arbitrarily compartmentalized humanities and technical learning. This split, as evidenced in the case of the University of Illinois at Chicago (UIC) and Illinois Institute of Technology (IIT), discouraged inquiry into the political features of technical practices, including issues of gender or race equity. When the university installed military historian Frank E. Vandiver as president in 1981, he inherited a situation with regard to minority opportunity and race relations at College Station that he believed to be the "worst of any in the nation." The school employed only 17 black faculty, and enrolled only 323 black and about 1,000 Hispanic students, far below proportionate representation of those minority groups in the state. Vandiver founded a Committee on Minorities to investigate and improve faculty and student conditions in regard to race; the committee soon issued unstintingly critical reports on the "political, social, academic and spiritual atmosphere" at TAMU as discouraging to minority recruitment and retention.[17]

TAMU leadership approached the problem of increasing the campus' minority population as a matter of both curricular change and the introduction of new services for students with diverse needs. Significantly, the school began to consider race as one of sixteen factors in its admissions decisions alongside gender, family income, SAT scores, legacy factors, and athletic participation. These measures quickly bore fruit. From a population of under 1,500 black and Hispanic students in 1981, that population reached 3,369 by 1987, with some reports measuring retention rates for those students as near 90 percent, far higher than the state average of around 64 percent.[18] In 1986, the school created a summer program to introduce local minority high school students to medical professions and to provide introductory coursework in a variety of fields.[19] TAMU's College of Engineering (named for alumnus H. Dwight Look in 1994) soon became the most successful division in this regard. New personnel, from the dean of Engineering downward, were hired with the expectation that they would recruit more minorities and enact pedagogical innovations that would adapt to different learning styles.[20]

From 1983 onward, when TAMU began efforts to enroll more minority engineering students, the college incorporated a program of improved recruiting and then expanded tutoring, counseling, and financial assistance. The following year the engineering college created a staff position, filled by Jeanne Rierson, to oversee these efforts. Initiatives for minority students grew still further when Herbert Richardson, formerly at MIT, was hired as the new dean of Engineering in 1985. In a pivotal decision for the future of minority inclusion efforts in the college, Richardson appointed Carl Erdman, who had served at TAMU as head of the Nuclear Engineering Department since 1981, as associate academic dean. Erdman undertook the increase of minority involvement in engineering as one of his primary responsibilities. Many of the innovations instituted under Erdman emphasized the importance of students sharing a sense of social identity beyond the traditional Aggie sensibility. Students were given multiple opportunities to self-select for programs that emphasized racial or gender commonalities.[21]

In its first manifestation, Erdman's concern with students' experience of community produced a small learning center shared by multiple engineering departments, in which minority students studied and socialized together. There is some evidence that within a year or two, TAMU's engineering departments saw a considerable increase in the number of bachelor's degrees granted to black and Hispanic students due to these innovations.[22] Many programs at TAMU offered minority students the chance to network "as much as they desire[d] with students from similar ethnic backgrounds." This priority led to the creation of special sections of some math or science courses expressly for African American or Hispanic students. Each of these sections was served by an upper-level undergraduate or a graduate student (usually him- or herself a member of an underrepresented minority) who would attend the class and then lead "supplemental instruction" sessions. For administrators, the goals of these sections included academic improvement for minority engineering students and "an organized approach to helping them find connections on campus."[23]

This agenda challenged Aggie traditions on campus that encouraged students to see themselves as members of the A&M community who shared values and experiences transcending their differing family or ethnic origins. Such traditions ostensibly promoted nonracial, communal impulses, but alumni and faculty have described the Aggie spirit as sometimes working against acceptance of cultural diversity on campus, especially when it has drawn on origin stories involving Confederate leaders such as Sul Ross. In the 1990s, a group of self-identified "Old Ags" objected to the erection of a statue commemorating Matthew Gaines, a former slave and state senator who had advocated for the creation of a black land-grant school in Texas in the 1870s. Even students who in 2007 supported the erection of such a statue did so because they believed that commemorating Gaines's contributions "would counter the mistake of requiring feckless multi-cultural courses inspired by the worst forms of political correctness."[24] In contrast to that kind of supposedly race-neutral ideology, minority education initiatives explicitly integrate knowledge work in classroom or laboratory with heightened experiences of students' ethnic identity. As Percy Pierre's leadership of PVAMU corroborates, students' minority heritage can become a foundation for social collectivity that in no way contravenes intellectual attainment, but in fact supports such attainment.

For educators like Erdman, emerging trends in STEM education funding encouraged exactly such integration. In 1990, the NSF inaugurated the Alliance for Minority Participation (AMP); building on its existing programs for minorities, the Texas A&M System became one of twenty-five AMP programs across the United States and Puerto Rico. Renamed the Louis Stokes AMP (LSAMP) in 1999, the program involved campuses in the Texas A&M System and nine community colleges, offering "an inviting academic and emotionally supportive minority student community." Interventions centered on bridge activities for high school or transfer students interested in engineering: scholarships and stipends, mentoring by peers, faculty and administrators, supplemental instruction, and special programs in connection with research projects at the university or at industrial sites. Minority students were clustered for coursework and related activities; "cohorted learning communities" were central to TAMU'S LSAMP efforts, to which the university system made a $50 million commitment. Actively embracing the goal of minority participation at all levels, many faculty in TAMU's engineering departments worked with the dean of sciences at Prairie View to bring its graduates to TAMU as doctoral students.[25]

But not all engineering faculty at TAMU accepted the premise that increased attention to minority participation held benefits for the institution, the profession, or for society at large. As one administrator recalls, some faculty objected to the provision of special support for students deemed to be disadvantaged, since they themselves had "made it" in the profession without any such help. Moreover, TAMU's drive to enroll more minorities was unfolding in a period when the university had many more applicants in engineering than it could admit, so the notion of expanding the pool of applicants held little appeal for those not otherwise committed to diversity. Some engineering instructors expressed a concern that when minority recruiting programs encouraged young people to choose engineering careers, those programs instilled in applicants a false sense of identification with the discipline: those youngsters who were "meant" to be engineers would already know that about themselves.[26] A study of LSAMP in 2002 by the Urban Institute, a nonprofit, nonpartisan policy organization, acknowledged that students needed to come into engineering training with certain characteristics already in place. That requisite familiarity with "professional performance and discourse," they stressed, is not achieved equally by all students: "researchers find that African American, Hispanic, and American Indian students and women of all races/ethnicities are less likely than white or Asian males to be exposed to experiences and opportunities that are the precursors to membership in a scientific community."[27] But the report does not propose that universities redress this inequity by reconfiguring notions of eligibility; LSAMP did not represent an attempt to correct racial inequity through that particular sort of change. A second NSF-funded initiative may have held greater promise than the LSAMP for directly upending these demographic patterns. The EEC, in which TAMU became involved in the 1990s, functioned within established accreditation structures (through ABET) and industrial or political patronage systems, but still managed to posit, if not achieve, some dramatic reforms to the social relations of engineering. Here, distributions of authority within the classroom came under scrutiny. With the EEC, the presumed individualism of engineering that left majority privilege unexamined at least received an airing, if not a thorough address, as a brief history of TAMU's participation in one EEC-sponsored coalition will show.

TAMU and the Foundation Coalition

The NSF began the EEC project in 1991 with a number of challenges to conventional academic ideologies. The agency's intention was "wide scale reform of engineering education" with associated inclusion of "underrepresented" groups such as women and minorities. As engineering studies scholar Juan Lucena has summarized this initiative: with the goal of a more flexible American workforce, one that is able to adjust rapidly to changes in the global manufacturing economy, the NSF sought to move engineering education beyond the scientific emphasis in place since the 1960s toward the preparation of a more mobile and practically oriented labor force.[28] This aim was wedded to the project of diversifying the U.S. workforce in terms of race, ethnicity, and gender. Each coalition would consist of approximately six or seven engineering schools or pre-engineering programs at different institutions. Providing $2 to $3 million over five years in each award, the EEC funded groupings of schools that included historically black and traditionally white universities, or that joined state universities, Ivy League schools, and small private polytechnics in a single initiative. Major tenets of the program included interdisciplinary connections among science, math, engineering, and humanities subdisciplines; "active" or hands-on learning with links to real-world conditions in which engineering would be applied; and a strong emphasis on assessment and dissemination of coalition members' activities. The program's linkage of intellectual and social change was patent. With each challenge to an existing intellectual category (such as its promotion of integrated course materials or its encouragement of student input regarding teaching styles), the EEC created institutional space for new distributions of credit, responsibility, and opportunity that would pave the way for social realignments within and beyond the university.

The NSF funded eight coalitions over the next ten years; most received a full decade of support. Some of the coalitions based their work on concrete goals regarding educational content, such as one of the first two groups to be funded, a unit formed by California Polytechnic State University-San Luis Obispo, the University of California at Berkeley, and Cornell, Hampton, Iowa State, Southern, Stanford, and Tuskegee Universities. That group, concerned with expanding the use of "advanced information technology" in teaching and learning, focused on courses in "'mechatronics', the merging of mechanical and electronic devices in contemporary engineering." By contrast, another early coalition to receive NSF funding (a unit that included Howard University, MIT, Morgan State, Penn State, the University of Maryland, and other schools) emphasized the role of design in undergraduate engineering education but gave centrality also to "the recruitment and retention of women and underrepresented minorities in engineering." A third coalition, based in the Detroit area, combined both agendas, featuring attention to the technical demands of the automobile industry and a pronounced focus on nontraditional students "often ignored by academia."[29]

By creating working groups of schools with different demographics, but more significantly with different levels of student performance and occupational aspiration, the NSF promoted a major change in the social profile of engineering education. In one sense, the NSF simply promoted the legitimization of pedagogical scholarship and the use of engineering faculty time and energy for something other than research, laying the groundwork for all sorts of progressive changes in the engineering classroom.[30] In another sense, if some federal funding patterns reinforced the elitism of top-tier schools by channeling research money disproportionately to a handful of universities, here a federal agency subverted structures that stood in the way of egalitarian social reforms.[31] By linking schools of different status, the coalition infrastructure undermined long-standing ideas about the origins and meaning of school reputation. Crucially, it conceptualized higher education as a collective societal project in which responsibility for educational opportunity (here, the chance for members of underrepresented groups to become engineers) would be shared among institutions. In a development that particularly pleased the NSF, members of one coalition worked collectively to help a single member school, a historically black university, to gain ABET accreditation for its newly established engineering program. It is likely that not all schools within a coalition carried equal influence or took away results of equal value. It would appear that in one case a historically black participating school with fewer engineering resources was able to adopt innovations of other schools, but not offer many of its own. In such cases the isolation of any single school would be systematically challenged by the NSF's coalition-based funding.[32]

In 1993, TAMU helped form the fifth EEC coalition to be funded by the NSF. That group consisted of the Texas A&M System's main campus at College Station and the system's predominantly Hispanic branch at Kingsville; Arizona State University (ASU); Texas Woman's University; the University of Alabama; the Maricopa Community College District (which served as a feeder system for Arizona State University's upper-division engineering programs); and the Rose-Hulman Institute of Technology in Terre Haute. Of these seven schools, TAMU, Arizona State University, and the University of Alabama were understood to have existing full-fledged graduate and research programs alongside their undergraduate engineering programs; Texas Woman's University and the community colleges offered pre-engineering options but not engineering degrees. The Foundation Coalition, as the group called itself, received $15 million from the NSF over five years, with coordination of the entire coalition undertaken by the Texas Engineering Experiment Station at College Station.[33]

The Foundation Coalition planned first to establish an innovative lower-division engineering curriculum, and then to redesign upper-division discipline-based coursework. The group would also create bridge programs to pre-college and community college programs to facilitate the transfer of those schools' students to baccalaureate programs in engineering. In 1994, as the group prepared to offer its first pilot courses for freshmen, Erdman summarized its work plan as promoting "continuous curricular improvement" and "increasing the number of degrees earned by currently under-represented groups." The integration of those goals may have reassured faculty who, as minority affairs staff at TAMU recall, felt far more comfortable with programs designed to enhance the learning of all engineering students, not solely those deemed to be for underrepresented groups.[34] But the linkage of pedagogical content and issues of minority access also represents a contrast to the standards-based approach seen in anti-affirmative action educational programs.

The depth of this progressive commitment is evident in two ways. First, in its ideas about altered course content, the Foundation Coalition's work reflects a pattern similar to the initial formation of engineering programs at UIC a decade earlier: the dismantling of conventional disciplinary boundaries in favor of integrated curricula. For example, instead of classroom work that presented math and physics concepts as entirely distinct topics, the new integrated freshman curriculum at TAMU would integrate those concepts across disciplines. Erdman offered as an example the idea that instead of seeing "three apparently unrelated" introductions to vectors, students would experience a single, integrated presentation of that concept as it applied across disciplines. The Foundation Coalition here followed a model developed at Rose-Hulman, which had restructured introductory courses in calculus, general chemistry, statics, mechanics, electricity and magnetism, and engineering design and graphics into three, first-year courses at twelve credits per quarter. This restructuring brought attention to the role of fundamental scientific principles in engineering problem solving and design. In Erdman's view, this sort of learning more closely imitated the kinds of multidisciplinary applications of knowledge that students would face in the "real world."[35] This imitation was crucial to the NSF, for whom systematic reform of undergraduate engineering curricula was meant to serve national economic interests. But it also allowed a loosening of cognitive ownership, in which disciplines violated conventional divisions of academic labor, newly sharing responsibility for instruction.

The coalition's leaders also reimagined interpersonal relations within engineering education and through that reform the possibility of increased participation by students who might otherwise have felt marginalized. Perhaps consciously, the Foundation Coalition planners gave this most overtly social feature of their work a distinctly technical name: "Human Interface Development," or HID. Under the HID rubric, faculty/faculty, student/faculty, and student/student interactions would all be recast in a mode of cooperative learning, engendering closer relationships at all levels. As he helped inaugurate the new Foundation curriculum, Erdman believed that increased communication and cooperation among these factions in and of itself increased the likelihood that traditionally underrepresented groups would "thrive" in engineering programs.[36] This claim stands in contrast to a model of professionalization based on the adaptation of the newcomer to existing modes of behavior, like that seen in some of the LSAMP evaluations, noted above. Erdman's acknowledgment of social milieu was also significant because it institutionalized attention to nontechnical matters in the course of instruction, not only through the provision of tutorial or psychological support services. In multiple ways, the Foundation Curriculum embedded social experiences as an integral element of technical education. The program's planners detailed the advantages of team problem solving paired with individual accountability, recasting notions of achievement in order to connect collective and individual learning. Teams were formed according to research on conditions conducive to the retention of women and minority students in higher education. Concepts were tested using both individual and team exercises, and lent (at least for the moment) equivalence to the two kinds of class work.[37]

In some ways, the Foundation Coalition's emphasis on "learning communities" echoed long-held priorities of MEPs that offered minority students "safe havens" from which to build students' academic confidence and skills.[38] But something more was going on here than the provision of shelter from the stresses of college life. Instead, ideas about social empowerment were being built into the engineering curriculum. In a further, potentially transformative redistribution of power in the engineering classroom, Foundation Coalition planners also credited students with a sort of meta-level understanding of learning processes. Using a set of "quality principles" throughout their first year, all students were expected to determine "the pedagogy that best fits their learning style." The Foundation Coalition planners institutionalized students' "expertise" through "Teaching and Research in the Undergraduate Experience," a joint project for students and instructors in curriculum development and research, and the "Student Intern Program" in which students visited other schools within the coalition to share their experiences. Foundation Coalition professors were "coaches" rather than instructors "or all-wise dispensers of knowledge." Extensive use of interactive technologies (such as software and video conferencing), and the elimination of lectures or recitation-oriented teaching, were central to moving students from passive to active learning. Lectures, if used at all, were to be limited to no more than ten minutes in duration, a startling departure from conventional formats.[39]

Those running the Foundation Coalition found evidence of its success at the end of its first year: of ninety-eight students who began the program, eighty-five remained in the program for the entire year. Foundation students showed academic gains in physics, calculus, and critical thinking that exceeded those of nonparticipating students. While in some other areas differences were not evident, observers agreed that Foundation students were better able to handle complex engineering problems and present their ideas in written form than were their counterparts. Most participating freshmen believed that the program was something they would choose again or recommend to friends. The majority of students reported that they had enjoyed being part of the teams and formed close relationships with instructors, as Foundation Coalition planners had hoped. TAMU decided to expand the pilot program for 1995–1996 to include 200 more students, and Arizona State University, another Foundation Coalition member, began its own program that year as well.[40]

And yet, for all these markers of change in curricular design and outcomes, the Foundation Coalition was disappointing to some. When NSF auditors approvingly found that student networks at TAMU and other Foundation Coalition schools were "well informed and highly judgmental" about different instructors, they signaled the importance of achieving social realignments to the EEC program. But when assessing all programs in the

EEC in 2000, the NSF found that the initiative had "fallen far short of desired systemic reform." At TAMU itself, in 2004, Dean of Faculties Karan Watson found that the numbers of women and minorities in engineering were stagnating.[41] A combination of structural and cultural intractability held Foundation Coalition reforms in check, and it is not always easy to see which of these causes might have engendered the other.

Understanding the Limits of Reform

Certainly, at TAMU, bureaucratic obstacles to pedagogical change were considerable. Changes on the classroom level required adjustments at the department and university level as faculty and staff were reassigned, classrooms redesigned, and equipment and supplies purchased and installed. Because EEC guidelines stressed the use of cutting-edge educational technology, many coalition schools faced the problem of keeping up with new developments in computers, including purchase and training. But if large-scale reforms seemed dauntingly complex, small projects struck some faculty and staff as causing more trouble than they were worth. Grants amounting to millions of dollars may seem generous on first consideration, but by the time that money is divided into myriad smaller projects and tasks, no one recipient receives very much. For TAMU departments other than those in engineering proper, including physics, mathematics, and chemistry, which were involved in the new curriculum but with less investment in immediate pedagogical gains, the entire project seemed prohibitive. Nor would all engineering departments perceive equal benefits through the achievement of coalition-inspired reforms. With its heavy programmatic emphasis on design, mechanical engineering seemed the primary beneficiary of EEC-funded changes at College Station, once it appeared that resources might have to be spread thin. On the most prosaic level, merely redistributing administrative responsibilities and staff work hours could cause problems. Transferability was a main goal of the NSF, anticipating ripple effects from its grants to coalition participants, but institutionalization of EEC innovations was hampered because the creation of teaching materials and routine administrative operations never achieved economies of scale.[42]

Whether these practical problems arose from, or caused, insufficient faculty buy-in to coalition reforms is difficult to assess. A number of EEC schools reported resistance to curricular reforms on the part of "old guard" faculty, who may have had entrenched interests in maintaining traditional curricular or classroom practices.[43] But there is little doubt that in most coalition schools, despite stated intentions to the contrary, reward structures simply did not change enough to allow even enthusiastic faculty a sense of security about spending their time on pedagogical projects. This was particularly pronounced at the research-intensive schools, although Watson believes that in the case of Foundation Coalition members, it may have been less of a problem at TAMU than at ASU or the University of Alabama. TAMU had expanded to graduate and research work more recently than those other coalition schools and thus had fewer established practices to upend. In any case, the NSF admitted in 2000 that traditional bases for tenure and promotion had altered only slightly in schools trying to implement the new engineering curricula.[44] For all its embrace of novel pedagogical content and new definitions of individual or institutional merit, the NSF's coalition project could not overcome embedded academic values in most settings. Coalition participants worried that once NSF funding ended their universities would revert to old curricular models, having failed to achieve structural change or enlist large and varied audiences.

Without support from the wider university community, even targeted minority programs in a single college or department might eventually face erosion. In the 1990s, a climate of general public acquiescence toward racial disparities on the campus could also be detected. Minority recruitment suffered as the parents of black students worried about "black themed" parties at TAMU fraternities. Racist cartoons published in the school paper periodically set off "right to publish" debates, and some students flew Confederate flags, which they claimed were protected by a "constitutional guarantee of free speech." One student group, in an effort to protest affirmative action policies, held a bake sale at which students of different ethnicities were charged different prices. In the surrounding community, black students were often mistaken for local residents when they stepped off campus, even decades after African Americans had first enrolled at College Station. A journalist who had long followed issues of race in Texas public universities suggested that the University of Texas attracted the state's minority students more readily than TAMU through its academic strengths. But the less diverse demographic at College Station and ongoing racial tensions may have led those students to enroll elsewhere, as well.[45]

Conservative political factions in Texas in these years seemed at best to tolerate, and at worst to facilitate, this racially fraught situation. Over the years, many plans were developed for minority recruitment at College Station. Enrollment figures for black and Hispanic students did rise, and those monitoring the situation acknowledged some progress, but a general feeling persisted among minority advocates that absolute numbers remained low. Outside observers of the efforts made by the Texas A&M System felt that, "[h]istory shows that nothing really happens when they don't reach their goals."[46] When George W. Bush became governor of Texas in 1994, allocations of public funding for Prairie View A&M and Texas Southern universities increased, especially for direct spending on students, but faculty salaries and facilities at the two historically black schools lagged far behind those at the College Station branch, the University of Texas, and the University of Houston. At TAMU after 1996, conservative interests included a state legislature eager to dismantle affirmative action and supported by a sweeping national movement in this direction. The reversal of legal protections against racial bias had been on the conservative agenda since the Supreme Court's *Bakke* decision of 1978. That decision had approved the use of race in college admissions if used in conjunction with other factors. In contrast, the *Hopwood* case formalized the sentiments of many conservative factions in and beyond Texas that any consideration of race was simply untenable.[47]

Cheryl Hopwood was one of four white students who claimed that the University of Texas had rejected their applications to its law school on the basis of race, accepting minority applicants with lower test scores. The Fifth Circuit Court of Appeals found in favor of the students, essentially outlawing considerations of race in admissions. The attorney general of Texas then applied that decision to all public colleges and universities in Texas. Furthermore, scholarships could no longer be offered on the basis of race. As predicted by the president of the University of Texas, the impact across higher education in Texas was immediate and dramatic. The number of minorities enrolled in science, math, and engineering in Texas dropped almost immediately. The number of minority doctorate-level degrees dropped as well, and by 1999 the pattern of minority undergraduate and graduate students leaving Texas for out-of-state schools that maintained favorable conditions for support and study was distinct. Dozens of out-of-state colleges, eager to recruit and offer scholarships to Texas minority students, set up recruiting operations to do so. At TAMU, enrollment of African American freshmen dropped 23 percent in the first year after *Hopwood*.[48]

In 1997, the Texas state legislature instituted a "top 10 percent" law, in which any student graduating in the top 10 percent of an accredited Texas high school would be guaranteed admission into a Texas public university of the student's choosing. For the University of Texas in Austin, the new system, combined with aggressive recruiting and increased financial aid packets, brought numbers of minority students back up toward pre-*Hopwood* levels, although it did not bring about a representation of black students that was proportional to their representation among Texas high school graduates. TAMU saw no adjustment whatsoever. The number of Hispanic freshmen enrolled at College Station dropped from 713 in 1996 to 570 in 1999. African American freshmen dropped from 230 in 1996 to 158 in 2003. At a point in the early 2000s, when African Americans made up 12.3 percent of Texas' college-aged population, blacks comprised only 2.4 percent of TAMU's entire enrollment.[49] Concerns that the top 10 percent system contradicted the traditional role of land-grant schools as gateways to higher education appear to have been well founded.[50] The state's minority "brain drain" accelerated as African American and Hispanic Texas families sent their children to more hospitable settings. In 2000, Texas remained one of three states, along with Virginia and Maryland, still found to be not in compliance with Title VI of the 1964 Civil Rights Act.[51]

Texas was not alone in this ambivalent stance regarding diversity. At the end of the twentieth century, Americans' concerns about minority disadvantages increasingly focused on the idea that racial categories were themselves the source of continuing educational and economic inequities. This notion, instantiated in President Bill Clinton's "One America" initiative on race, wedded rhetoric about fairness and democracy to the elimination of minority set-asides and affirmative action plans. As political scientist Claire Jean Kim has aptly summarized these measures, Clinton firmly positioned national unity as an achievable goal through multicultural dialog, but cast genuine recognition of and attention to race as sources of past conflict. This ideology found a foothold in TAMU's operations when the university installed Robert Gates as president in 2002. Gates announced that minority enrollment would be a personal goal, and soon after arriving he created the post of Vice President and Associate Provost for Institutional Diversity. His aggressive minority recruiting scheme involved the creation of eight "regional prospective student centers," and then the omission of information about race from applications.[52] New scholarships, distributed on the basis of merit and income, facilitated the recruitment of significant numbers of African American and Hispanic Texans. According to Dean of Engineering W. Kemble Bennett, enrollment of black students in engineering rose 92 percent under Gates's plan.[53] But these numbers reflect a correction of extremely low black and Hispanic enrollments prior to this period. One Texas senator noted that the institution "had a chance to correct its miserable record in admitting minorities and it did not take full advantage." More importantly, Gates offered a system that would eventually plateau as the small segment of minority Texans' who satisfied conventional measures of eligibility were recruited. Despite the findings of a 2000 internal report that TAMU was "an enclave for the education of White students," Gates offered solutions that reinforced achievement metrics that had long supported those racial differentials.[54]

Gates's elimination of legacy admissions was a strong gesture in the face of the potent Aggie traditions often passed down from one generation to the next.[55] In his willingness to address entrenched cultural habits at College Station, Gates as president has been compared to Earl Rudder. Gates is seen to have successfully begun a move for TAMU away from its image as a setting constrained by tradition and rearguard social outlooks. However, he did this by appealing to the pronounced meritocratic sensibility that scholars identify as characteristic of the post-civil rights period.[56] Thus, when the U.S. Supreme Court gave its approval to some use of race in college admissions in 2004, Gates did not reestablish race as a criterion for admission at TAMU. The University of Texas did so, but Gates believed that "personal merit" must remain central. In the new "race-blind" system, admissible students would be offered $5,000-a-year scholarships if they came from families earning under $40,000 a year, and if they were the first in their families to attend college. The top 10 percent system remained in place, but Gates believed that the new scholarships would reward students on the basis of grades, leadership, work experience, and, it would appear, some generalized notion of fortitude. In many ways, his objections to affirmative action confirmed ideologies of knowledge-work as an enterprise without social and political meaning, while actually deeply implicating social differences. His conception of TAMU's culture was one "grounded in patriotism, religious belief (however expressed), loyalty to family and one another, a hard work ethic [and] character and integrity."[57] These are culturally specific priorities that privilege very particular ideas of high moral character.

According to those responsible for minority participation in TAMU's engineering departments after 2000, sentiments that arose from engineering faculty helped reinforce the idea that a focus on merit assured fairness. The College of Engineering operated on the premise that it should admit more students than it could keep, on the presumption that a certain proportion would drop out. Difference in skills and talents, which comprise a sort of "natural selection," must be allowed to operate, and enhancing retention seemed no more pertinent than minority recruitment. Some engineering faculty were willing to provide special opportunities for students of lower socioeconomic background whom they perceived to have been disadvantaged, but not for students identified on the basis of race or gender. But even to distinguish among those categories of experience seems an arbitrary choice, little influenced by empirical evidence about the larger societal function of race in American higher education.

As Jeanne Rierson noted in surveying TAMU's programs for minority engineers, "There will never be enough data to convince people about this . . . you have to have committed leadership, like Carl Erdman."[58] But even with strong, progressive leadership, academic engineering programs generally play out in a larger climate in which considerations of equity are compartmentalized from bench-level training and practice. By 2003, TAMU's top leadership had delivered its clear endorsement of race-blind academic culture. It seems reasonable to conclude that the university was no longer a setting in which an overtly reformist program regarding race in higher education could readily pervade engineering. Even those Foundation Coalition innovations that challenged existing pedagogical patterns in the most obvious way—by supporting greater involvement by students in curricular design or by developing communal learning situations—had to stop short of systematically scrutinizing race, gender, or other self-understandings central to modern social relations. In this somewhat discouraging climate, Foundation Coalition initiatives, in order to survive, maintained some of the firewall that had stood between technical knowledge production and political activism for generations. As had many programs for minority students in engineering since the 1970s, these efforts acknowledged experiential features of technical pedagogy, and at their best, the importance of flexible standards and learning styles in higher education, but could not address the most entrenched social structures contributing to racial discrimination.[59]

NASA at Prairie View

It takes little more than manifestations of the Aggie spirit at TAMU, in which images drawn from a historically homogeneous and white demographic united much of the huge student body, to remind us that technical spheres express larger cultural values. If Gates's concessions to racial tension at College Station (including inaugurating "Soul Food Fridays") seem small scale, and unlikely to dislodge existing patterns of white cultural dominance, in many ways he expressed a national mindset.[60] In this climate of reluctance regarding the open address of race at the end of the twentieth century, the racialized identities of HBCUs presented a number of conundrums for educators and their patrons. Here, race could not go unnamed, but black empowerment could be circumscribed or diluted. Legal requirements for increased white enrollment at public HBCUs, including the Office of Civil Rights' instructions to PVAMU, were seen by some educators as ironically requiring both a renewed focus on race and a denial of historical inequities from which the HBCUs arose. A progressive impulse within the federal establishment, born in the early days of the civil rights movement, seemed to counter that conservative trend by assuring research funding from the nation's top science agencies, specifically to HBCUs. Under Carter's Executive Order No. 12320 of 1980, U.S. Departments of Energy, Defense, and Health and Human Services, as well as the Environmental Protection Agency and other science- and non-science-centered bodies, earmarked a portion of funding for HBCUs. That program, intended to "strengthen and expand" HBCUs, built on ad-hoc efforts of various federal agencies of the previous decade. It brought considerable funding from NASA to many minority institutions and continues to expand in education and research areas today, now including in its purview Hispanic- and Native American-serving colleges and universities. It has been renewed by every president since Carter, and supported by the White House Initiative on HBCUs created under Reagan.[61]

To conclude the series of case studies that have served as our window on race in U.S. engineering education, we return to PVAMU where, beginning in the early 1990s, materials engineers and scientists received funding from NASA that ultimately created a vital center for research with space and civilian applications. In this program and its successful implementation at PVAMU, the association of the school with a black identity formed a basis for funding. A series of initiatives at NASA through the 2000s reiterated this approach, as one set of NASA program managers put it, "obliterating myths" that had severely limited HBCU research opportunities in the heavily funded universe of space science research.[62]

With its origins residing in U.S. anxiety about Soviet scientific achievement, NASA had operated educational programs almost from its inception in 1958, with concerted outreach to K-12 and college-age students.[63] An impetus toward increasing minority involvement at the graduate and research level at NASA arose from Carter's requirement that federal agencies assign a growing proportion of their research budgets to minority institutions of higher learning. Dozens of small, historically underfunded university science and engineering departments, some with virtually no previous history of research or even of graduate instruction, began receiving research funding from the nation's highest-tech agencies. Schools that had never received measurable federal science funding before, some of which had only rudimentary laboratories and had produced no patents, were newly able to share in appropriations previously reserved for universities of far higher profile in science and engineering. After a century of pronounced fiscal disadvantage, these disenfranchised institutions might now hope for fuller participation in national research and development agendas. Science grants to black universities and colleges remained a relatively small outlet for the federal government; in 1991, 104 HBCUs received $700 million from a pool of over $11 billion awarded to all U.S. colleges and universities. However, significant scientific and engineering contributions, based on unique expertise in both teaching and research, issued from the federally supported HBCUs.[64]

This scenario provides all the requisite elements for vigorous debate regarding affirmative action in the sciences.[65] There are challenges to conventional criteria for funding in higher education in order to offset historical race-based inequities; an assertion of racial identity not only as a meaningful but as a central category in determinations of practitioners' scientific worth; and a centralized federal body distributing largess to geographically dispersed recipients, sometimes overriding local decision-making authorities. Each practice has been characterized by opponents of affirmative action as morally objectionable and undemocratic as well as a potential threat to the quality of American scholarship. Without the mechanisms of "pure" competition, unhindered by social planning, there is no guarantee that the best scientific minds will find support and reward.[66] According to this outlook, some presumptive nonminority contributions are being lost as opportunities are *denied* to some practitioners on the basis of their race. Such voices have been gaining in volume for the last two decades. The White House Initiative on HBCUs remains unusual in spite of documented achievements by HBCU science and engineering programs functioning under its auspices. Moreover, it counters a persistent impression that most HBCUs would "naturally" hew to their founding missions of helping black Americans achieve a modicum of economic mobility rather than strive to produce and disseminate knowledge for its own sake, as one group of analysts recently summarized this retrograde impression.[67] On the broadest level, secure federal funding for HBCUs represents a rare confluence of national agendas for technological advancement and for social reform, with something to tell us about the historically limited compatibility of those agendas in the American imagination.

Radiation Research at Prairie View

The College Station and Prairie View branches of the Texas A&M System were founded through land-grant legislation in the 1870s that proposed to harness higher education for the purposes of scientific and general economic expansion. As described in the previous chapter, however, PVAMU, like many other "colored" branches of state land-grant university systems, was for many generations a far smaller and drastically underfinanced enterprise than its majority-white counterpart. The school, having moved beyond its strictly agricultural and mechanical origins by the 1920s, focused for several decades on readying students for all vocations in which black Texans were likely to find employment: teaching, nursing, and selected industrial areas for which engineering technology degrees were appropriate. While TAMU formed the Texas A&M Research Foundation in 1944, and the Texas Engineering Experiment Station (founded in 1914) continued to grow, engineering degree programs inaugurated at PVAMU in the 1940s only expanded to graduate-level teaching in the 1960s, with a single master's program in general engineering. A decade after the end of legal segregation in 1964, the school held little hope of participation in the region's growing research and development sector through graduates' employment or research.[68]

In contrast, the predominantly white College Station campus has been a nationally important site for aeronautical and aerospace, petrochemical, computing, and other areas of military and industrial research since midcentury, when it expanded beyond its own early agricultural and mechanical functions. Growth of the Texas economy and of the state's predominantly white public universities (which also include the 48,000-student University of Texas System) have long been linked, and Texas A&M and UT have had tremendous influence on the creation of high-tech industries in the state, in exchange receiving steady funding in technical areas.[69] But, as described in the previous chapter, Prairie View and Texas Southern University—the state's other public HBCU—were by and large outside this funding structure until 1980, when Texas' black legislators prevailed in their efforts to redirect portions of oil-derived public higher-education funds to the minority campuses.[70]

PVAMU's leaders immediately applied these resources to new degree programs, new physical facilities, and a recasting of the school's role in regional engineering efforts. In 1982, with the new state funding in place and indications of significant research commitments from the federal government, PVAMU planned a major upgrading across its technical divisions. As president in this period, Percy Pierre cultivated close relationships between the university and Bell Labs, IBM, General Motors, and other companies with research and development interests. Through what were in some cases long-standing corporate faculty-loan programs (such as those set up by IBM and AT&T), personnel from those firms spent lengthy visits teaching at Prairie View and inaugurating research projects.

This activity paved the way for additional master's degrees in civil and then electrical and chemical engineering by decade's end.[71]

At this juncture, 90 percent of university science and engineering research in the state took place at TAMU and UT, and circular but influential arguments supported the idea of restricting developments among the state's other public institutions when NASA, already in established relationships with the Texas' majority universities (including TAMU), began supporting research and teaching at the HBCUs under the federal program.[72] PVAMU Dean of Engineering Austin Greaux and a group of PVAMU alumni met with Hans Mark, director of NASA's Ames Research Center. That visit began a process that led to site visits at Prairie View by Mark (eventually promoted to deputy administrator at NASA) and Aaron Cohen, director of the Johnson Space Center (JSC) in Houston. As the closest NASA operation to Prairie View, JSC offered a ready outlet for students and graduates of PVAMU seeking experience with the space sciences. Like other NASA centers and laboratories, JSC undertook a major effort to diversify its personnel on the basis of race, national origin, religion, and gender. It supported a program for minority graduate researchers and hoped to increase the number of scholars eligible for that program through the provision of direct aid and advice to HBCUs, including PVAMU. The federal HBCU program spurred a still larger commitment in terms of research funding, and NASA inaugurated its own wide-ranging, cross-disciplinary program of support to HBCUs. This program included the creation of seven University Research Centers (URCs) beginning in 1991, expanding to fourteen by the end of the decade and including, in that second set of URCs, the Center for Applied Radiation Research (CARR) at PVAMU.[73]

From the start of NASA's funding there in the early 1980s, PVAMU's most visible space-related work had centered on applied radiation research, "an understanding of space radiation effects on electronics and biosystems." In an interdisciplinary arrangement that included chemistry, electrical engineering, and mechanical engineering faculties, the university established a research agreement with NASA for the study of charged particles, an area crucial for NASA's Space Shuttle, Space Station, and Hubble Telescope projects. PVAMU's "Five Task Research Agreement" with the agency addressed system-level effects of radiation; many applications of researchers' findings to non-space projects, such as the development of solar-cell materials, nanofabrication, and metal-oxide semiconductor technologies, rapidly became clear. Representatives from IBM, McDonnell Douglas, and other corporations worked closely with PVAMU faculty in this period to support the new research programs. PVAMU brought increasing numbers of its physics, biosystems, and agricultural faculties to work on these projects until in 1992 the work transcended departmental divisions. With the goal of becoming a "national center for research on radiation science, detection, shielding and tolerance," the university that year established the Laboratory for Radiation Studies. That unit began receiving NASA funding of about $1 million per year, becoming the Center for Applied Radiation Research (CARR) in 1995.[74]

Between 1994 and 2004, CARR received nearly $5 million from NASA for projects associated with multiple research, flight, and space centers, and the Jet Propulsion Laboratory.[75] CARR's strengths spanned radiation research from the atomic level up to integrated systems. Under the initial direction of Thomas Fogerty, an electrical engineer who had worked on radiation effects for defense contracts at AT&T before moving to Prairie View, the center took as its motto, "Radiation Safety from Earth to Space." Human/technology interfaces, dosage measurement, and investigations of testing processes brought new collaborators, including TAMU's College of Engineering and Stanford University, with PVAMU often representing the leading partner in these efforts. CARR's choice of research topics was linked to "expressed needs" of NASA's "Strategic Enterprises," which included its Human Exploration and Development of Space Enterprise and its Aerospace Technology Enterprise, embedding the PVAMU personnel in a well-established research organization.

CARR's utility to research efforts grew beyond NASA. By the end of the 1990s, Los Alamos was soliciting CARR researchers to help with its work in neutron science. In 2000, when NASA designated PVAMU as a URC, CARR began receiving funding in five-year increments. Importantly, CARR was also producing increasing numbers of researchers drawn from the minority population. Bachelor's and master's students in PVAMU's engineering programs now regularly moved outward through involvement with CARR projects to doctoral programs at other universities. PVAMU's first Ph.D. program, in electrical engineering, developed within CARR. The training of African American students and other minorities "to feed into the Ph.D. pipeline" had been part of the original mission of the Laboratory for Radiation Studies, and continues to be a major goal of CARR today. Its directors see that demographic role as inseparable from the perpetuation of the unit and its research agendas.[76] This was never meant to be an advanced research enterprise that would flourish in spite of PVAMU's racial heritage, but in service to that identity. NASA was fully complicit in that project; the minority-institution URCs were jointly supported by the agency's Office of Equal Opportunity Programs and its substantive divisions, the Strategic Enterprises. It is in that intentionality that the full political character of CARR's operation becomes apparent.

The Meanings of Merit

By 2000, NASA's total "investment in HBCUs" had reached an annual amount of around $57 million. In that year, these funds went to forty-four minority-serving institutions (which included predominantly Hispanic and Native American, as well as African American, schools) in the form of 206 separate awards to faculty, students, and administrators. This translated, by the agency's accounting, into more than 292 refereed publications, 6 patents, and 227 science and engineering degrees granted to student researchers (including 11 doctorates). At this juncture, NASA was spending more than one billion dollars a year in funding to institutions of higher learning overall, and its allocation of roughly 5 percent of its funding for HBCUs is not wildly out of proportion with those schools' representation (at 3 percent) among all colleges and universities in the nation. This budget matched levels recommended by government advisors for HBCU-dedicated funding. However, closer scrutiny reveals that HBCUs matriculate close to 25 percent of all African American students attending four-year institutions. In that regard, NASA actually followed convention in earmarking a disproportionately small amount of its resources for minority opportunity. Yet NASA broke with convention to open science and engineering research to minorities, and considerations of scale alone may not be enough to measure the agency's reformist intentions and the potential cultural influence of its interventions in minority science opportunity. Inherent in NASA's interest in supporting minority institutions was its understanding that the provision of money alone would not suffice to assure HBCU participation in the technical sector. On at least two levels, NASA recognized that long-standing inequities had left the minority-serving institutions ill-equipped to compete for science and engineering funding and that without some foundational changes to structures of eligibility in science, inclusion was unlikely.[77]

On one level, to overcome historic conditions that had left the HBCUs outside of many university research networks, NASA promoted its Minority University Space Interdisciplinary Network starting in the early 1990s. This program involved the provision of support for technical development (including the creation of a computer networking infrastructure for Internet, distance-learning, and other purposes) through the designation of certain schools as Network Resource and Training Sites (NRTS). In this program, seven universities, including PVAMU, served as regional hubs through which additional two- and four-year institutions would attain connectivity and receive technical support.[78] NASA saw the NRTS as a way to expand eligibility for participation in its research and educational programs: "NRTS's are institutions that, by Cooperative Agreement design, had not historically received major amounts of funding from NASA, but could convince merit reviewers that they could implement the NRTS vision."[79] What is striking here is the explicit address of the inadequacy of existing merit structures and the commitment to new ones aimed at altering patterns of inclusion. What is more, in other instances, including competitions for minority-earmarked Department of Defense funding, HBCUs had to form their own networks before applying for funding. In this case, by contrast, NASA made the creation of such networks part of its purview. HBCU researchers understood that professional connections within university spheres were vital, and they effectively used the NRTS program for this purpose.[80]

An even more profound example of this kind of social openness and proactive stance is seen in NASA's embrace of what can only be regarded as an epistemic shift as it sought to bring long-excluded populations into play as competitive researchers. As part of a long-term plan to create a sustainable research infrastructure at PVAMU, NASA allowed CARR to produce usable results at a pace which was measurably slower than that of laboratories in majority institutions. According to one director of CARR, without that accommodation, which he calls "a combination of spending and patience," the center would never have been able to embark on its initial research plan, let alone to have achieved results while building up its facilities and personnel from scratch. He distinguishes that altered set of performance criteria from a "lowered bar." Like all NASA URCs, the center was subject to yearly peer review by NASA personnel. CARR's research outcomes, apart from the pace at which they were achieved, have been indistinguishable from those of competitors operating under conventional time constraints. Throughout its written policies regarding involvement with minority-serving institutions, NASA reiterates its intentions to prepare faculty and students at those institutions to "successfully participate in the conventional, competitive" research process.[81] Flexible time frames, which accommodate differentials in institutions' economic histories, pose no threat to the quality of research produced with NASA funding.

From the perspective of prevailing policies in engineering higher education, this severance of pace from other metrics of productivity was a fairly radical move on NASA's part. In a departure from common practice in the provision of federal science or engineering research, it achieved inclusion by acknowledging structural inequities that had hampered past performance and treating contemporary practice—possibly manifesting a slower rate of production—as an understandable byproduct of those inequities. For these historically underdeveloped research sites, NASA bypassed the customary eligibility patterns by decoupling two features of engineering work: the pace of laboratory production (for which expectations were adjusted) and the utility and reliability of the laboratory products (for which standards remained unchanged). This is not a disaggregation that historically has had wide application. Most programs intended to boost minority participation in science, technology, engineering, and mathematics do just the opposite, aggregating timing and performance. For example, as described in earlier chapters, in engineering teaching there has been an enduring rejection of remedial coursework. This rejection dates back to the first MEPs in the 1970s. It took hold on the basis that such coursework would increase the amount of time students would have to spend in college, and thus the cost of engineering education, but it would also lower the prestige of schools that provided remedial instruction.[82] That argument promotes an arbitrary recipe for the achievement of rigorous engineering in which timing is seen invariably to matter.

According to the staff of CARR, liberal developments at NASA came down to selected individuals within NASA who carried their own progressive values into their work for the agency with PVAMU. Not all potential funders expressed the kind of flexibility and concern that CARR saw operating at NASA, and even some NASA employees did not approach the HBCU project with identifiably inclusive intentions.[83] But it seems possible that the federal agency offered a milieu in which progressive individuals could tolerate the risks associated with maintaining flexible performance standards. Unlike academic settings, in which individuals and institutions might be derided for deviating from normal research protocols, the bureaucratic nature of the federal agencies may have allowed such departures. In such settings, employers might determine employees' value to the parent enterprise with a distinct calculus. NASA had a strong educational mission from its inception, and perhaps engineering research undertaken with the HBCUs fell under that special, "permissive" rubric. If conventional reward structures hindered the success of programs like TAMU's Foundation Coalition reforms, NASA's alternative priorities allowed research conducted at places like PVAMU to achieve first-rate status over time as engineering process and product were distinguished from one another in a way that promoted new social arrangements.

Were the creative and flexible program administrators who allowed CARR its lengthened "start up" time anomalous among NASA personnel? Or, did this inclusive program forecast broad social change, a model for programs through which the discriminatory mechanisms of American science might find correction? Many scholars have asked about the relationship of science and democracy in the United States, with a focus on

NASA as one particularly significant expression of national scientific and political values. They have asked since the 1960s about the relationship of NASA's overall mission—the allocation of national resources to military and industrial development—to the possibility of meaningful social change in the United States. Many critics over the last fifty years have seen federal expenditures on the space program as wasteful and misguided in light of the nation's profound social problems, including such racially marked issues as poverty, education, and decaying urban infrastructure.[84] What is more, the two arenas rarely conjoin in the American imagination. Neither government policies nor political rhetoric customarily project a world in which U.S. technical prowess is dependent on, or is a route to, the correction of social inequities. Jennifer Light described the efforts of the aerospace industry to become involved in solving "urban problems" in the 1960s, but at the highest level of federal administration, the two were seen as mutually exclusive priorities. What are we to make of Caspar Weinberger's admonishment of Richard Nixon regarding the president's neglect of the space shuttle program? In 1971, Weinberger urged Nixon to move forward with the program because to do otherwise would be to "give up our super-power status." The United States, Weinberger added, "should be able to afford something besides increased welfare, programs to repair our cities, or Appalachian relief and the like." Certainly Reagan, under whom Weinberger served as secretary of defense, sought to erode governmental commitments to social welfare programs and to promote the more militaristic applications of space science.[85]

But how then to explain that the federal impetus toward racial inclusion was, in this instance of NASA's funding of HBCUs, strong enough to alter epistemic habits that in other settings served to limit minority opportunities? NASA's extensive involvement with HBCU research perhaps implies a democratic domestic role for the space agency, even as it forwarded some of the country's most imperialist agendas. Certainly utopian ideologies put forth by many who supported the U.S. space program held that space travel and colonization represented a "clean slate" in which earthly problems might be left behind, as new resources and a frontier mentality promised "peace and justice." Less fantastic than such "astro-futurism," perhaps, were arguments like those made by Lyndon Johnson that associated the U.S. space program with geopolitical power and prestige: "In the eyes of the world, first in space means first, period; second in space means second in everything."[86]

In this worldview, the "culture of competence" that the Apollo program projected to the nation in the 1960s and 1970s, and, many presumed, to a global audience, was compatible with civil rights agendas. As much rhetoric of the period implied, a rising techno-scientific tide would raise all boats, and HBCUs would become part of a national landscape of research achievement deriving from the space program. CARR, under the leadership first of Fogerty, and subsequently Richard Wilkins and Kelvin Kirby, has undertaken work ranging from relatively narrow investigations of technologies to be used only in space (such as proton tests on International Space Station optoelectronics) to much broader research on commercial, off-the-shelf circuitry. Other minority institutions, including NASA URCs and collaborators with the URCs, produced findings of similarly wide industrial utility in bioscience, mechanics, photonics, and materials, among many other fields. There can be little doubt that NASA research funding was meant to yield not only new technical knowledge, but some cultural capital for the institutions, minority and otherwise, to which it flowed.[87] Certainly, the federal bureaucracy might have sheltered reformist thinkers in ways that more cut-throat industrial or academic research settings might not. Nonetheless, it seems most likely that NASA's commitment to HBCUs arose from a genuinely inclusive social vision (which extended scientific opportunity to minority Americans) that also sought to uphold the agency's basic economic and geopolitical character, and thus, that of the nation.

In essence, NASA's support of research at HBCUs expressed a belief that technical expansion and progressive social change were, at the very least, compatible and, at best, linked developments. I return to the question of how compatible these two goals actually are in this book's concluding chapter. For the moment, it will suffice to say that NASA itself clearly believed in that compatibility as it established an ever-larger bureaucratic apparatus for carrying research funding and educational initiatives to minority constituencies. Through such reformist reasoning, entrenched epistemic habits in science and engineering research became subject to revision. With those revisions, newly inclusive policies for science were established. It is the reach of such policies that needs to be measured. In 2000, NASA saw its involvement with diversity programs as a means toward a "more diverse resource pool" of "science and technology personnel" to "confront technological challenges for the benefit of NASA and the Nation."[88] In the precise nature of that "benefit" lays the political meaning of the federal research program for HBCUs, as this book's conclusion will investigate.

Conclusions

To summarize, while many of the innovations inaugurated by TAMU's Foundation Coalition project remain in place today, and new programs for minorities in engineering take shape at College Station through the efforts of extremely committed faculty and administrators, the essential top-down character of this pedagogy persists. Reward systems for academic engineers continue to put a premium on research, thus assuring not only that pedagogical projects operate on only a limited scale but also that established hierarchies of prestige, with heavily funded research scholars at the top, preclude any far-reaching role for students in shaping curricula or classroom practice. The "learning community" model, which promised flexibility and thus inclusion in Foundation Coalition courses, can have only a limited impact in a professional culture in which non-expert voices carry so little weight. This pattern disempowers all students, but particularly discourages disadvantaged students who might have found a welcome in engineering through such open social models. Overall, as affirmative action remains a peripheral strategy, at best, for the TAMU system even after Gates's departure, the recruitment policies put in place under his administration seem to satisfy many of those currently responsible for bringing diversity to the university.

Meanwhile, at PVAMU, research sponsored by NASA continues to bring prestige and new funders to the university from corporate and government sectors. In 2008, NASA provided Wilkins, Kirby, and other researchers with a $5 million grant to establish the school's next generation of radiation research facility, the Center for Radiation Engineering and Science for Space Exploration, or CRESSE. As interest among U.S. college students in spacecraft engineering diminishes, the enlistment of women and minority students into these fields gains new urgency and justifies inclusion initiatives on all of the Texas A&M System campuses.[89] And yet, NASA's flexible ideas regarding funding, which did open important high-level research opportunities to historically marginalized institutions like the HBCUs, still function in service to the relatively inflexible economic hierarchies on which U.S. science and engineering are grounded. This may in part explain why NASA's HBCU program has not become a widely followed model for STEM interventions.

Ultimately, undergraduate instruction in the Foundation Coalition scheme had to "look enough like" conventional engineering to reassure doubters; HBCU laboratories funded by NASA had to fit "strategic goals" of that agency to offset charges of marginality. In both cases, concern about racial equity could be derailed by manufactured anxieties about the lowered standards or altered technological priorities that many fear accompany inclusion. With these two cases in mind, we need to return to the prospect that eligibility for full participation in U.S. science and engineering arenas pivots on ideas of merit that are socially inflected, and that (as sociologist David K. Brown has explained) credential standardization brings "positional advantages in bureaucratic and professional labor markets."[90] To put it slightly differently, engineering reflects Brown's general finding about higher education that a discipline's need for organizational certainty engenders uniformity in training and professional practice. This is why dramatic but relatively small-scale diversification outcomes (such as increased black enrollments achieved at TAMU and the impressive research growth of CARR at PVAMU) cannot be considered likely to have unlimited ripple effects: their proliferation would represent too great a divergence from customary institutional approaches.

Identifying a fear of conformity hardly provides a satisfying historical explanation for conformity, but there is something to be gained by probing the stakes that engineering schools may see in diversification efforts. The potential of these programs for replication at other sites of engineering teaching and research is dependent not only on the development of new institutional commitments, but on thoughtful analysis of what engineering believes it may risk in undertaking such social extensions. Most important, membership in engineering spheres is not infinitely expandable, as we might expect of an arena in which detailed technical knowledge is necessary, but it is in fact severely constrained by factors having little to do with the fitness of aspiring practitioners. For one thing, such expansion is simply impractical in science and engineering as it currently functions because, as historian Stephen Shapin points out in his description of "core sets" in science, it is through constrained membership that scientific communities hold onto their unified social identity: "it is predominantly the members of this small group of familiar others who hold the immediate fate of one's knowledge-claims in their hands."[91] That is, without limits, membership cannot function as such. If Shapin is correct, the Foundation Coalition's community-based learning and NASA's funding for HBCUs bring about constructive changes, and welcome African American and other minority practitioners into engineering, but subsist partly because they do not project unpredictable or systemic change. Rather, they project limited-scale inclusion (and here I refer to limits of both numbers of participants and research directions).

The question arises, then, about how susceptible engineering might be to a greatly expanded conception of membership. In other words, to exactly what problem is the creation of a more racially diversified engineering profession—through MEPs, funding set-asides, and even affirmative action programs—meant to be the solution? As we near the end of this study, we can aim to understand just what a "diversified engineering workforce" means to those for whom it is a priority. In few cases is that goal considered to be the achievement of a more equitable or democratic society, in any simple sense. More often, *diversity* is associated with an increased labor pool or access to a variety of problem-solving approaches, somehow embodied in women or people of color as groups thought likely to enhance America's global technical performance. On a deeper level, the pursuit of a diversified workforce may hold more portentous meaning, skewing social interventions into academic science and engineering toward those programs that maintain existing opportunity structures, even while admitting increasing numbers of minority citizens into technical occupations. For all their authentic concern about minority experiences in higher education, MEPs will not achieve greater impacts until we acknowledge their potential to serve these conservative social goals.

CHAPTER EIGHT

Conclusion

Engineers are problem solvers.
—*Fundamentals of Engineering,*
Mark Holtzapple and W. Dan Reece, 2003

IN EXAMINING THE HISTORY of black underrepresentation in U.S. engineering, I have taken on the task of describing absences. This means that there are not always prominent individuals or seminal events that might neatly delimit the narrative. Instead, the focus is often on actions not taken and on social and cultural developments that failed to come about. When historical actors decide not to act, that too represents a choice. When American educators and policy makers after 1945 chose not to maximize opportunities for minority students, even when presented with legal precedents, or later, with their colleagues' innovative diversity plans, they were both expressing personal ideology and responding to cultural conditions. The delineation of historical absences shows how those forces combined in the second half of the twentieth century to constrain black participation in engineering. It may also point the way to alternative paths for the future.

One absence considered in this book is that of African American students from U.S. engineering degree programs. The low number of black engineering students persists despite three generations' efforts to promote racial equality in higher education. Another absence is that of open discussion of identity in science and engineering fields. The effacement of identity has marginalized talk about engineers' sense of themselves and others in relation to race, gender, ethnicity, and nationality. These two absences have supported one another over the last fifty years. Thus, by extension, in acknowledging the historical role of identity in scientific and technical practice we can hope to achieve greater diversity. For engineers' identity to become visible as a factor in patterns of educational and career attainment, it must step out from behind practicality, the signature feature of engineering in many professional self-descriptions.[1] Once visible, identity may figure in a reform of inequitable patterns of STEM participation.

The epigraph of this chapter is taken from the opening pages of an intriguing recent textbook for incoming engineering students, a book that provides a clue to how we might go about this disclosure. That textbook, unlike many, frames its technical content with a lengthy discussion of social matters. These issues range from environmental sustainability to the challenges of protecting manufacturing jobs in the face of global outsourcing. The authors include "a few words on diversity," warning students that impulses toward "tribalism" and "overgeneralizations" about coworkers of different background will stand in the way of productive and respectful teamwork. Sidebars highlight achievements of notable female and African American engineers, bringing those often-overlooked individuals to students' attention. However, in at least one way the textbook obeys convention: in its depictions, the engineer's professional conduct remains distinct from his or her ethnicity or gender. The reader is shown an occupation devoted to achieving practical aims through disciplined workplace acculturation. We might build on the textbook's suggestive message of social concern and its heterogeneous image of engineering knowledge work to, instead, foreground identity.[2]

The textbook's messages about engineering are both descriptive and prescriptive, and already mesh the social and the intellectual in an exciting way. In the authors' telling, individual traits such as "logical thinking," "follow-through," "curiosity," and "common sense" enable the professional engineer to achieve design aims of "simplicity, increased reliability . . . smaller size, lighter weight, etc." Yet, we read, successful practitioners also systematically put aside individual differences in the interest of collective technical goals. Good engineering also requires the cultivation of ethical and moral sensibilities, to which an entire chapter is devoted. But as the textbook is written, race and gender remain ancillary to meritorious engineering work. Neither category of self-identity is routinely to be addressed as a determinant of social experience or acknowledged as part of the student's intellectual apparatus. The authors' advice to students on workplace relations appropriately includes respect of ethnic difference: the avoidance of ethnic jokes because people should "do unto others as they would have done unto them." But identity is not actually described as a potentially constructive intellectual element, as an essential source of knowledge that pertains to engineering.[3]

This absence of race or gender identity in the science or engineering classroom is no less important for being conventional, as the cases in the preceding chapters have shown. It is an elision that leaves the predominance of certain groups in the field unmarked. Why not expand the engineering textbook's social project by integrating students' background and life experiences into their technical work? Why not make social relations a *primary* feature of rigorous engineering practice? This would likely encourage the retention of minority and female STEM students, as diversity experts have long suggested.[4] But it would also help achieve the acculturation aims already outlined in the textbook. The race consciousness of the minority student might engage with complex social matters and ethical concerns; the majority student might routinely see the specificity of his or her life circumstances and those of others. All students might develop a sense of the social origins of the "economics" with which, the authors tell us, engineers must contend.[5] Moreover, with this explicit address of social structures, engineering students might understand themselves to be not only problem solvers, but problem shapers; that is, people who choose to undertake some tasks and not others, with consequences for the wider world. Recognizing the junctures at which choices are being made is antecedent to making more informed choices, a goal for student engineers that the textbook authors clearly support.

There is little doubt that this kind of change is more easily described than implemented. Explicit attention to minority identity has customarily been countenanced in U.S. engineering and science only when accepted standards of technical conduct appear to be quarantined from social intervention. Opponents of race-based educational programming and affirmative action initiatives in engineering have commonly drawn on anxieties about lowered standards; as early as the 1960s critics distinguished between the "fairness" of some programs for disadvantaged citizens and regrettable "free rides" offered by others (see Chapter 4). That view of standards as encapsulated guarantors of rigor posits some features of engineering practice as being devoid of political intentionality—uninflected with any notions of self-interest or cultural difference. But there is no level of conduct within engineering that can ever be below or before the social. Science studies scholars understand that honest scientific practice is as fully social a matter as fraudulent practice. Similarly, we can see that definitions of rigorous engineering practice invariably express social ideologies, including those having to do with race and social equity more generally. A historical perspective emphasizes the durability of patterns of racial, ethnic, and gender representation in STEM fields. It also encourages us to see that those categories of social experience, those ways of knowing ourselves and others as members of a professional set, as Steven Shapin might put it, do not cease at the classroom or laboratory door.[6]

In doing so we see that some conceptions of engineering talent and rigor used to support segregated engineering programs at the University of Maryland in the 1940s, and others used many decades later to deny the necessity for race-based interventions in Texas in the 1990s, invoke common ideas about technical work. *Racism* is not necessarily a label that teaches us much about these exclusionary patterns, because intentionality may have varied widely among the educators involved. But across these fifty or sixty years, the demographic consequences of such educational policies have remained the same and we can interrogate educational practices on that basis, while helpfully identifying exceptions to these patterns. Enlarging the social messages that a textbook might give freshmen engineering students is one way of seeing that the history of race in U.S. engineering education has something useful to teach us; its telling is not a purely retrospective exercise meant to yield moments of regret or self-congratulation about America's approach to racial equity. The following pages provide four general areas in which the historical cases featured in this book might suggest further constructive changes.

1. *Expand our ideas about the sorts of character traits and educational background that may produce a successful engineer.* As a number of episodes in this book have made clear, engineering educators have historically treated students as "input," destined to arrive at the end of their engineering training as "output" or simply to fail to make that entire journey through the university. Almost by definition, every university engineering program expects that desirable candidates will have arrived at college age with either sufficient academic skill to undertake the standard coursework in engineering or observable potential to attain those skills. The problem, as John Brooks Slaughter reminded us in NACME's recent report on minority STEM participation, is that this sort of "observable potential" by and large remains a very narrow category.[7] Few students who have failed to present conventional attainments (say, those with low SAT scores) appear to be good risks; even effective initiatives such as NACME's Vanguard Engineering Scholarship Program remain small as the engineering and educational establishments continue to function within such conventions.

Instead, prevailing standards of high school achievement that now discourage many minority students from pursuing technical careers could be loosened (as UIC attempted in the mid-1960s). Remedial college coursework or lengthened curricula could be provided for those students who need these services, a manipulation of pacing like that undertaken in

NASA's research programs at PVAMU. Formal learning communities, such as those established at Texas A&M's College of Engineering under the Foundation Coalition program, could become the rule rather than the exception. Such collective experiences that put students and instructors on equal footing for selected purposes are shown to encourage retention of so-called underprepared students. They also helpfully reconfigure the judgments of faculty about students' potential; such programs need not remain unusual. New scholarships could be established to serve this widened pool of candidates, likely attracting more middle- and working-class applicants who might not otherwise see themselves as able to attend college. Students who previously felt they had perhaps no other option than to join the military to obtain technical job training might instead enroll in academic engineering programs to pursue research careers. All of this requires money, but the aim of this book is to show that the nation's claimed high valuation of diversity in the engineering workforce calls for that fiscal commitment.

By extension, the nation's failure to see such spending as worthwhile reflects a limited commitment to equity. There can be no question about the conservatism of many applications of diversity in the United States since the 1980s. As Claire Jean Kim points out, multiculturalism in some forms fails to expose *vertical* differences among different categories of citizens. For example, President Clinton's "One America" initiative can be seen as "a high-profile example of how superficial multiculturalist rhetoric could be used to simultaneously grant minorities symbolic recognition and distract attention from the impact of racism on their material existence."[8] Some universities go beyond that simplistic and obfuscatory approach to diversity to seek a deeper range of values among their students. Arthur Johnson, a professor of bioengineering in UMD's Clark School of Engineering, and Rosemary Parker, director of the school's Center for Minorities in Science and Engineering, assessed the results of a cultural diversity survey at that university in 2001. Johnson and Parker have come to worry that "the faces of students in a class may look different, but once they open their mouths, their voices all sound the same." Importantly, these authors note that "selecting students with the highest SAT scores" defeats the ostensible purpose of diversity programs to bring variable social and political ideologies into the university. Even more crucially, the authors recognize the disincentives surrounding a university's departure from established academic standards. Within the academy, high-achieving students are universally perceived to be "easy to teach" and likely to become "objects of pride" for their instructors and schools. As we have seen, this was certainly the case among many faculty and administrators at IIT, TAMU, and many other prominent engineering schools when they maintained stringent standards of student eligibility. Concerns about institutional reputation that we saw operating in many of the schools described in this book tend to reassert the importance of such student characteristics over and over. But Johnson and Parker are drawing attention to factors that have remained unspoken in many schools, including their own employer, UMD, in earlier periods. Almost uniquely, these authors convey that "*the price paid for extreme selectivity* is a constriction of cultural diversity" (emphasis added).[9]

Their view is rare in proposing that higher education give institutional diversity precedence *over* selectivity. Diversity would thus cease to be a "special function" distinct from the fundamental operations of technical pedagogy, as minority education specialist Bryan Brayboy found to be the norm in U.S. universities.[10] Similarly, Shirley Ann Jackson, physicist and president of Rensselaer Polytechnic Institute, invokes the experiences of women and minorities in science and engineering in the following appropriately stark terms: "They have the talent, it only needs to be engaged, encouraged, nurtured and prepared."[11] She captures here one important process through which the underrepresentation of minorities in the sciences has come to be naturalized in U.S. culture. She emphasizes that while we accept that majority youth will need training and intellectual development before their economic contributions will be possible, many of these formative functions of education are denied to women and minority students. Thus, crucially, Jackson helps us see the arbitrariness of differentiations between qualified and qualifiable students so common in MEP operations. Rather than continue a long tradition in educational policy of treating students as input, we can probe the conditions under which student identity is shaped and constrained after arrival at the university. We should take seriously Johnson and Parker's identification of some widely established diversity practices in U.S. universities as representing no less than a "dishonest diversity" and "a fraud."[12]

2. *Reconfigure reward systems, top to bottom.* Universities will not adopt expanded notions of student eligibility for engineering degrees merely because a number of faculty and administrators see their value. As I have tried to make clear, engineering education functions as a system in which individuals strive to meet institutional expectations. At every level, from instructors up through trustees and directors, well-meaning people in the academy have faced disincentives to change. Among the cases presented here, frustrations faced by reformist thinkers at UIC in the 1960s and 1970s express this perhaps most clearly. The College of Engineering at

UIC eliminated some innovative programs for minority engineering students when those programs began to preclude conventional measures of success, such as research funding, and then to erode faculty retention. The case studies show that for many very understandable reasons, individuals and entire institutions currently shy away from work that might lead to greater minority participation in STEM degree programs. Reputations and rankings matter for the survival of academic careers and schools, so activities that seem to loosen standards (for incoming students, or for faculty research attainments) present intolerable risks.

We need to change that situation by systematically rewarding those activities that support demographic change. University heads such as Freeman Hrabowski, at the University of Maryland, Baltimore County, have done so within their universities to great effect. But it is also an area in which organizations responsible for reputations and rankings may make a huge contribution. As a clearing house for ideas about engineering education, the American Society for Engineering Education (ASEE) can help with this sort of culture change by tracking effective programs and offering new terms for our measurement of gender or racial diversity. Some very innovative work has emerged from ASEE members who use that forum to develop potentially radical ideas.[13] However, accreditation and funding bodies (ABET, NSF) too must alter their reward systems for universities and researchers if these ideas are to reach application.

On one level, this alteration could involve integrating racial, ethnic, gender, or other areas of student diversification into official metrics of institutional performance. Accreditation bodies might define objectives and desired outcomes for university degree programs to include the entry of underrepresented groups into STEM fields, rather than leave it to universities to determine the scale and priority of such matters. This reform need not take the form of numerical benchmarks, a notoriously difficult sort of policy to implement in an anti-affirmative action climate. Instead, ABET might require innovations in recruitment, teaching, or retention along the lines of those initiatives described above. Student demographics could thus be solidly integrated into accreditation structures, rather than remain a second order goal after conventional curricular standards and learning outcomes. This would obviously lend increased importance to undergraduate education, with graduate programs becoming a less powerful marker of university prestige. But as diversity expert Daryl Chubin has pointed out in blunt and compelling language, lamenting the size or diversity of the scientific talent pool while "feeding an insatiable research funding model that sacrifices all else to the production of knowledge" is "not only contradictory, but silly and stupid."[14]

A second constructive change to the reward structures under which engineering schools function might follow the lines of the NSF's Engineering Education Coalitions (EEC) program, which linked the activities of large, established engineering schools with smaller, less prominent ones. NSF funding in this program (described in Chapter 7) is based on coalition-level accomplishments. For example, EEC funding granted to Texas A&M and Rose-Hulman was contingent on collaborative work on teaching innovations. This sort of linkage disrupts the often self-reinforcing nature of science funding structures in which only those schools that have already seen success appear eligible for further support. This is a fairly radical idea because it appears to violate the function of standards to pinpoint individual failures and attainments, institution by institution. If we put aside the circularity implied in that argument, we can see that there is no reason collective practice cannot be judged as meticulously as any individual effort. This sort of collectivity needs to be extended to many more sites of educational practice if we are to imagine genuinely expanded patterns of inclusion.

Finally, academic reward systems could shift to make student diversification a central element of all engineering faculty positions. As Brayboy has indicated, administrators in many colleges and universities treat the pursuit of diversity as a free-standing project meant to engage faculty of color with students of color to create a more "user friendly" setting. This arrangement leads to a compartmentalized and thus marginalized address of race in the university, connoting that majority culture bears no responsibility for the construction of racial ideologies and stands to gain nothing from altering those ideologies. It also carries significant workload consequences for faculty of color, who face "implicit and hidden requirements" to meet their employers' diversity goals, while also facing demands on their time to undertake research and publication required for tenure and promotion. Thus, placed in the hands of overburdened faculty, a shrunken inquiry into race is substituted for one that could engage with far larger social patterns and opportunity structures. But this system of tasks and rewards has an even more insidious feature: because the overtaxed faculty have difficulty achieving competitive levels of research, their performances may corroborate lowered expectations about their scholarly capacities.[15] In each of these areas, universities need to reexamine the institutional structures within which their diversity programming is expected to function.

3. Demonstrate that authentic diversity can lead to new and sustainable markets for technical knowledge and products. Many of the educators and university administrators described in this book thought carefully about where their students would find employment, and reasonably found many of the kinds of changes I have just outlined to be unacceptably risky. Unconventional educational standards and curricular content that might have encouraged minority participation in engineering have retained their problematic aura for decades. In contrast to those standards, American corporations since the 1960s have consistently been among those most avidly trying to expand the representation of women and minorities in technical fields by encouraging and sponsoring their inclusion in many academic programs. It is worth thinking about this disconnect. How is it that industry, while eager for a diverse workforce, has not managed more effectively to empower the academy to provide that workforce?

As pointed out in Chapter 1, diversification arguments based on national workforce needs do not encourage minority inclusion to the same degree as broader, social justice arguments. Practical, economic concerns that dominate corporate logic by definition do not maximize the possibilities for full-scale social reform. But rather than pushing corporations away from practical concerns, it is necessary to make clear to them that widened eligibility for engineering professions can translate into material advantages for the production sector. That is, if members of diverse economic communities participate in the technical decisions of industries and governments, the interests of a wider range of consumers will be represented in those decisions. Markets for the goods and services produced in Western, technology-based industries may thus grow. This might help corporations see the value of greater support for minority participation in engineering: increased funding, more active recruitment, and endorsement of the reforms listed above.

This is not an essentialist claim along the lines that white, male engineers will tend to build interstates and invent nano-materials to meet current market demands, while female engineers or engineers of color will build mass-transit systems and design solar ovens to serve developing markets. That sort of facile association of practitioners' race or gender with certain intellectual proclivities or political ideologies is of course deeply problematic (although it has appeared in academic and corporate diversity literature since the genre's inception). However, the engineers discussed in this book are treated as people with meaningful life experiences, family heritages, and political commitments. University STEM programs currently select for students with a narrow set of educational experiences, and because education and economic background are so closely linked in the United States, true diversity in those three sets of characteristics is lost. Americans hold a much wider range of priorities than is now represented in engineering, and one can argue that such diversity is not valuable only for the sake of abstract humanistic principles. Rather, I would offer that the needs of non-affluent rural or urban communities or developing nations can translate into markets for businesses. An economically (and therefore, in the U.S. case at least, ethnically) diverse technical workforce may lead to expanded understanding of what counts as rational design, production, and marketing to different audiences. Some so-called green innovation and production, which envisions redistributions of material resources and wastes, already addresses corporate opportunities of this type. Ideally, the government might support engineering entrepreneurship and education with these inclusive goals in mind, but even without such progressive state reforms new markets are emerging that can justify expanded notions of who might contribute to technical work, and how.

There is a real risk in this suggestion of supporting cultural stereotypes and reinforcing the marginalized status of certain groups. We cannot deny that the engineering of cities, industrial processes, and energy infrastructures has historically disadvantaged some groups because it tends to serve existing distributions of wealth and power. Here we return to some of the largest and most challenging questions raised by a socially inflected history of engineering: Does today's lack of racial diversity in technical occupations reflect a leftover sort of discrimination soluble through educational reforms, or a truly intractable commitment to a racially stratified society? Merely pointing to the economic benefits of a diverse workforce will not reveal the cultural root-system that maintains inequitable opportunities for black citizens. In 2001, William Wulf, then president of the National Academy of Engineering, said that more women and minorities must be enlisted into engineering careers "because insufficient diversity means lower quality engineering!" (punctuation in original). He went on to specify that the "range of design options considered in a team lacking in diversity will be smaller," and that "the product that serves a broader international customer base, or a segment of the nation's melting pot, or our handicapped, may not be found."[16] But in this formulation Wulf, in many instances an energetic advocate of minority engineering programming, perhaps mischaracterizes the nature of patronage systems in which engineering operates. It is not necessarily the absence of diverse personnel that has led to the social narrowness of research agendas in engineering. Rather, it may be the narrowness of social interests on the part of those who design engineering curricula, determine student eligibility, fund research, and employ engineers that has led to the lack of diversity.

What is more, as social and cultural theorist Avery Gordon notes, a corporation may seek an ethnically diverse workforce simply in hopes of creating a public image of pluralistic management in the globalizing economy, and universities can inadvertently support that project. TAMU's description of outreach programs to Hispanic students is representative in projecting common national interests in a way that elides class and race differences: "If we could get the fraction of minority students going into space systems engineering up to the same level as that of non-minorities, then we'd be taking a huge step toward solving the national problem."[17] Surely there is not some underlying plan here to disadvantage minorities; quite the contrary. But it is on the level of such general coinage (the construal of the missing space engineers as a "shortage," let alone as a "crisis") that criticality toward engineering projects or distributions of national resources disappear. Often, national engineering interests are congruent with the interests of non-engineering elites: those who set wages, plan infrastructures, and garner profits from the use of public and natural resources. As the division between the most wealthy and least wealthy strata of the U.S. population widens, and corporations increasingly determine public policies regarding such matters as employment and environmental regulation, ignoring the material disadvantages associated with minority status becomes increasingly problematic. This is not to say that space science research is without value, or that every dollar currently spent on space research should instead be spent on projects of more immediate and general benefit to more citizens. Those questions remain open. But my goal is to point out that in reductively constructing a sense of common national purpose around existing science or engineering goals, we help to hide differential conditions faced by majority and minority populations even in times of an expanding economy.[18]

4. Remind all those involved in engineering teaching, research, and employment that social change historically has had a place in this discipline. All three of the preceding suggestions have involved flipping our usual priorities to put social outcomes above, or at least on equal footing with, the material and economic functions of technical work. This would imply that social outcomes are customarily excised from conversations about engineering. And yet, historically, engineers have claimed social influence and reform as part of their professional work. Some of the language used in Holtzapple and Reece's opening textbook pages about engineers' social and ethical responsibilities would not have been out of place in a textbook of a hundred years ago. What has been lost in many discussions of diversity in engineering is the possibility of deep self-critique that might maximize the possibilities for reform.

In some historical contexts, the claim of a social function for engineering has actually reinforced structures of occupational and economic privilege. As historian Jonson Miller has shown, for generations following the creation of American engineering as a distinct professional sphere in the nineteenth century, a shared identity emerged as educated, white males encouraged fellow engineers to maintain an exclusive profile for the occupation. This continues today to some degree. If occupational identity represents a form of trust among insiders, as sociologists tell us, those who are not trusted can never achieve occupational status. Heather Dryburgh, in her study of female experiences in engineering programs, sees the cultivation of "solidarity" among engineering students as serving the development of "professionalism."[19] But, at the same time, those shared aspirations of professionalism elicit feelings of social commonality for students. Presumptions of common goals and a common likelihood of success help engineering students press forward through their studies, even if collective effort itself is not the norm. Once graduates enter the workplace, employment confers a further sense of belonging to those who have achieved professional status. Whether those feelings of occupational connection will trump a sense of advantage among majority practitioners to make black, Hispanic, Native American, physically challenged, or gay and lesbian persons trusted colleagues is not yet clear. Clearly, occupational participation in STEM fields, like almost every other professional realm, need not be a zero-sum game as some anti-affirmative action arguments seem to propose. Because professions depend on preserving a sense of commonality among their members, however, existing feelings of professional privilege readily perpetuate themselves. Certainly, as Alice Pawley, an engineer who has studied the experiences of women in the discipline, articulates, if the social advantages currently accruing to majority students and professional engineers continue to go unmarked, such changes are less likely to come about.[20]

It seems likely that relatively few educators would agree with commentator Peter Wood, who recently wrote in the *Chronicle of Higher Education* that, "a society that worries itself about what chromosomes scientists have isn't a society that takes science education seriously."[21] Yet we might still wish to imagine a new kind of question about the role of race in higher technical education or professional engineering practices; one that directs our attention to the content of these practices as an instrument for controlling eligibility. In each era since the late 1960s, a large percentage of reformist educators have condemned the underrepresentation of blacks in engineering as an unintentional byproduct of a previous era's lapses. Alternatively, reformers have displaced responsibility for low minority representation at the professional level from universities onto high schools or public school systems in general. College and university personnel cast their institutions as standard bearers for high-level knowledge, or, more rarely, explicitly as entities trapped in systems of status and reputation that preclude the sorts of radical curricular transformations that might encourage minority involvement. This study recommends that we recognize the social instrumentality of rigor as another important and persistent, if unintentional, means of perpetuating African American absences from STEM fields.

AS BARACK OBAMA'S ELECTION has shown, many people in America today are comfortable with the idea of ending the nation's historical connection of whiteness and social authority. Many who voted for Obama surely did so because that change felt long overdue. As the nation moves toward a "minority-majority" population in the new century, retrograde race-based notions about opportunity and aptitude will face new challenges, but many people have hoped and worked for progressive educational reforms for decades, long before that demographic shift was on the horizon. Among engineering educators of the last fifty years we find some of the most vigorous advocates of this kind of change, forward-thinking and self-sacrificing people who have laid out ambitious plans for diversifying a huge occupational field. How is it that such good intention and so many good ideas have resulted in relatively limited change? That question cannot be answered by looking only at engineering education; we need to look at American understandings of race and difference overall. Kim has warned that multiculturalism as currently configured in America addresses racial difference in a way that operates "at no cost" to officials or powerful groups in U.S. society.[22] For example, it does not include the financial costs associated with a lengthened college curriculum that adds remedial coursework for graduates of weaker high schools. Nor does it carry the social and economic costs that would come with relinquishing occupational privilege, should established professional cohorts (white, or male) face a more open occupational field, operating with more flexible definitions of merit. The claims among universities, corporations, and policy makers to care about minority inclusion in engineering are many, but the disincentives to true diversification in the sciences are still more numerous. There are few settings today where aims of racial diversity upend the basic tenets of stratified labor that we associate with our modern industrial society to shift resources in the form of taxes or public services (including education) from the most advantaged to the most needy sectors of American society. It remains to be seen if these shifts will emerge under Obama's administration.[23]

Nonetheless, it is important to avoid a concluding sense of the glass being half-full or half-empty; either feeling can lead to inaction as we become either complacent or despairing about racial equity in engineering. There certainly are both optimistic and pessimistic voices on which to draw. On the one hand, historian V. P. Franklin points to the strong cultural role being played by black schools today and their proven potential to empower minority students. On the other hand, William Watkins reminds us that black education historically served the interests of a deeply racist capitalist system.[24] Perhaps the most constructive way to view relations between American engineering education and corporate capitalism is not as a determined one, but as an enduring one. The durability of conservative ideologies in engineering and their power to limit the impacts of more progressive conceptions about economic and intellectual opportunities in the United States arise from specific institutional processes. These occur within engineering departments, schools, and professional contexts, as well as on the broader level of educational policy formation and legislative activity. So too, these processes play out on the level of admissions standards, technical curricula, classroom exercises, textbooks, tests, and instruments, where a role for race may seem most unlikely. When we recognize that technical knowledge itself has such social uses, we may begin to see the myriad possibilities for changing those uses.

Notes

Index

Notes

Abbreviations

BIHE	*Black Issues in Higher Education*
CHE	*Chronicle of Higher Education*
CHM	Chicago History Museum
IITA	Illinois Institute of Technology Archives, Paul V. Galvin Library, Chicago
JBHE	*Journal of Blacks in Higher Education*
JBS	*Journal of Black Studies*
JNE	*Journal of Negro Education*
PVAMUA	Prairie View A&M University Archives, Prairie View, Texas
TAMUA	Texas A&M University Archives, College Station, Texas
UICA	University of Illinois at Chicago Archives
UMDA	University of Maryland Archives, College Park

Preface

1. Mike Nizza, "After the Jena 6 Case, a Spate of Noose Incidents," *New York Times*, October 10, 2007; Peter Applebome, "Racial Crisis? Or Just Rope in the Hands of Fools?" *New York Times*, October 14, 2007; "NYC Declares Day against Hate Crimes," *New York Times*, November 1, 2007.
2. NACME, "NACME 2005 Data Book" at www.nacme.org (accessed 8/11/2007).
3. One of the few descriptions of such an experience that I have seen in writing is by Philip J. Sakimoto and Jeffrey D. Rosendahl, "Obliterating Myths About

Minority Institutions," *Physics Today* 58, no. 9 (2005): 49–50. See also Agyeman Boateng, "Viceroys of Difference: Minority Engineering Program Directors at Mid-Atlantic Universities" (Master's Thesis, Drexel University, 2007).

4. Charles Hamilton, "The Place of the Black College in the Human Rights Struggle," *Negro Digest* (September 1967) cited in Samuel DuBois Cook, "The Socio-Ethical Role and Responsibility of the Black-College Graduate," in *Black Colleges in America: Challenge, Development, Survival,* ed. Charles V. Willie and Ronald R. Edmonds (New York: Teachers College Press, 1978), 57. Cook describes Hamilton's "militant" advocacy of self-accreditation or "perhaps no accreditation" as flawed, and characterizes it as a "recipe for disaster for the black college and the black-college graduate" (pp. 56–57). This tension between divergent outlooks regarding black participation in mainstream science and engineering research requires historicization.

5. David Stout, "Justices Rule for White Firefighters in Bias Case," *New York Times,* June 29, 2009.

1. Introduction

1. Lisa M. Frehill and Nicole M. DiFabio, "The Status of African Americans in Engineering," in NACME, *Confronting the "New" American Dilemma: Underrepresented Minorities in Engineering: A Data-Based Look at Diversity* (White Plains, NY: NACME, 2008), 23.

2. NACME, "Patterns in the Production of Minority Engineers, 1994–1995," *NACME Research Letter* 6, no. 1 (1997); NACME, "NACME 2005 Data Book" at www.nacme.org (accessed 8/1/2007); NSF, Division of Science Resources Statistics, "S & E Degrees: 1966–2004" at www.nsf.gov (accessed 8/6/2007); Eleanor Babco, *Underrepresented Minorities in Engineering: A Progress Report* (Washington, DC: American Association for the Advancement of Science, 2001), 7; John Brooks Slaughter, "The 'New' American Dilemma," in NACME, *Confronting The "New" American Dilemma,* 7.

3. In addition to NACME's publications, and statistics issued by the Commission on Professionals in Science and Technology and the National Academies of Sciences and Engineering, helpful recent overviews of minority participation in STEM fields include Bryan McKinley Jones Brayboy, "The Implementation of Diversity in Predominantly White Colleges and Universities," *JBS* 34, no. 1 (2003):72-86; Shirley M. Malcom, Daryl E. Chubin, and Jolene K. Jesse, *Standing Our Ground: A Guidebook for STEM Educators in the Post-Michigan Era* (Washington, DC: American Association for the Advancement of Science, 2004); Marta Tienda et al., *Closing the Gap? Admissions and Enrollments at the Texas Public Flagships before and after Affirmative Action* (Woodrow Wilson School of Public and International Affairs, Princeton University, 2003). For a recent example of how some universities reconcile affirmative action policies and selective admissions, see Heather Schwedel, "Study: Affirmative Action's Job Far from Done," *Daily Pennsylvanian,* October 11, 2006, 1, 4.

4. Bruce Seely, "Research, Education and Science in American Engineering Colleges: 1900–1960," *Technology and Culture* 34, no. 2 (1993): 344–386.

5. Elaine Seymour and Nancy Hewitt, *Talking about Leaving* (Boulder, CO: Westview Press, 1997).
6. NSF, Division of Science Resources Statistics, "S & E Degrees: 1966–2004"; "Black College Students Overwhelmingly Pursue Business Degrees," *JBHE* 13 (Autumn, 1996): 22–23.
7. NACME, "Patterns in the Production of Minority Engineers"; NACME, "NACME 2005 Data Book"; NACME, *Confronting the "New" American Dilemma"*; NSF, Division of Science Resources Statistics, "S & E Degrees: 1966–2004."
8. W. G. Bowen and D. Bok, *The Shape of the River: Long-Term Consequences of Considering Race in College and University Admissions* (Princeton, NJ: Princeton University Press, 1998).
9. Juan C. Lucena, *Defending the Nation: U.S. Policymaking to Create Scientists and Engineers from Sputnik to the "War against Terrorism"* (Lanham, MD: University Press of America, 2005).
10. Peter Wood, "How Our Culture Keeps Students Out of Science," *Chronicle of Higher Education* 54, no. 48 (2008): A56. For extensive critiques of such claims, see Christopher Newfield, *Unmaking the Public University* (Cambridge, MA: Harvard University Press, 2008) and Donna Riley, *Engineering and Social Justice* (San Rafael, CA: Morgan & Claypool, 2008).
11. Slaughter, "The 'New' American Dilemma," in NACME, *Confronting the "New" American Dilemma*," 7.
12. Newfield, *Unmaking the Public University*, 92–106. On the reductive nature of such measures, see Joseph Farrell, "Equality of Education: A Half-Century of Comparative Evidence Seen from a New Millennium," in *Comparative Education: The Dialectic of the Global and the Local*, ed. Robert F. Arnove and Carlos Alberto Torres (Lanham, MD: Rowman & Littlefield, 2003), 154; Daryl E. Chubin, "Transcending the Places That Hold Us: Public Policy and Participation in Science" (Paper presented at Workshop 2000: A Joint Conference of the American Association for the Advancement of Science, the Emerge Alliance, and the National Science Foundation, Atlanta, GA, February 24, 2000).
13. In highlighting lasting inequities in engineering and locating the source of those inequities in the content and goals of modern technical disciplines, this study is intended to augment, not displace, accounts of pioneering individuals and institutions. Important examples of such accounts include Portia James, *The Real McCoy: African-American Invention and Innovation, 1619–1930* (Washington, DC: Smithsonian Books, 1989); Rayvon Fouche, *Black Inventors in the Age of Segregation: Granville T. Woods, Lewis H. Latimer, and Shelby J. Davidson* (Baltimore, MD: Johns Hopkins University Press, 2005); and David Wharton, *A Struggle Worthy of Note: The Engineering and Technological Education of Black Americans* (Westport, CT: Greenwood Press, 1992).
14. The book builds on analyses such as those recently offered by John Carson and Christopher Newfield regarding the social instrumentality of merit more generally. Those authors compellingly describe the ways in which beliefs about the nature of students' innate intellectual capacity—how to measure, compare, or predict individuals' or groups' likely educational attainments—have shifted as

different political ideologies have gained influence in Western cultures. Newfield traces the powerful role that prevailing metrics of educational ability have played in furthering minority disadvantage in the U.S. universities into the twenty-first century, and this study carries this inquiry into the corridors of engineering. Christopher Newfield, *Ivy and Industry: Business and the Making of the American University, 1880–1980* (Durham: Duke University Press, 2003); John Carson, *The Measure of Merit: Talents, Intelligence, and Inequality in the French and American Republics, 1750–1940* (Princeton, NJ: Princeton University, 2007); Newfield, *Unmaking the Public University.*

15. "Negroes Open Fight to Enter South's Colleges," New York *Herald Tribune,* August 25, 1935.
16. On the persistence of biological understandings of race and intelligence in the United States, see Troy Duster, *Backdoor to Eugenics,* 2nd ed. (New York: Routledge, 2003). William H. Watkins, *The White Architects of Black Education: Ideology and Power in America, 1865–1954* (New York: Teachers College Press, 2001), 39.
17. Joe R. Feagin, Hernan Vera, and Nikitah Imani, *The Agony of Education: Black Students at White Colleges and Universities* (New York: Routledge, 1996); x–xii, 1–13.
18. Paul Burka, "Agent of Change," *Texas Monthly* (November 2006): 155–159, 258–264.
19. Thomas Gieryn, "Boundaries of Science," in *Handbook of Science and Technology Studies,* ed. Sheila Jasanoff et al. (Thousand Oaks, CA: Sage Publications, 1995), 404–415; Michael Omi and Howard Winant, *Racial Formation in the United States,* 2d ed. (New York: Routledge, 1994), 10–11; Steven Shapin, *A Social History of Truth: Civility and Science in Seventeenth-Century England* (Chicago: University of Chicago Press, 1994); Michael F. D. Young, "An Approach to the Study of Curricula as Socially Organized Knowledge," in *Knowledge and Control,* ed. Michael F. D. Young (London: Collier-Macmillan, 1971), 19–46. To understand how engineering enacts cultural agendas common to many or all academic disciplines, we can consult the literature on diversity and expertise in capitalist cultures more generally. See Avery Gordon, "The Work of Corporate Culture: Diversity Management," *Social Text* 13, no. 3 (1995): 3–30; Jane Juffer, "The Limits of Culture," *Neplanta* 2, no. 2 (2001): 265–273; Manning Marable, *How Capitalism Underdeveloped Black America* (Cambridge, MA: South End Press, Updated Edition, 2000); and Newfield, *Ivy and Industry.*
20. Such needs for knowledge workers can be historicized, as sociologists of education have shown since the 1950s, and thus help explain how a stratified labor scene sustains itself. See, for example, Randall Collins, "Functional and Conflict Theories of Educational Stratification," *American Sociological Review* 36 (1971): 1002–1019; Raymond Murphy, "Power and Autonomy in the Sociology of Education," *Theory and Society* 11, no. 2 (1982): 179–203. In addition, see Basil Bernstein, "On the Classification and Framing of Educational Knowledge" (pp. 47–69); Nell Keddie, "Classroom Knowledge" (pp. 133–160); and Michael F. D. Young, "An Approach to the Study of Curricula."

21. James D. Anderson, *The Education of Blacks in the South: 1869–1935* (Chapel Hill, NC: University of North Carolina Press, 1988), 1, cited in John Hardin Best, "Education in the Forming of the American South," in *Essays in Twentieth-Century Southern Education: Exceptionalism and Its Limits,* ed. Wayne J. Urban (New York: Routledge, 1999), 9.

2. Identity and Uplift

1. William Collins, "Race, Roosevelt and Wartime Production: Fair Employment in World War Two Labor Markets," *The American Economic Review* 91, no. 1 (2001): 272–286.
2. George H. Callcott, *A History of the University of Maryland* (Baltimore: Maryland Historical Society, 1966), 364.
3. Four states (Mississippi, Virginia, South Carolina, and Kentucky) had set aside funds from the first Morrill Act for black public higher education. Oscar James Chapman, "A Historical Study of Negro Land-Grant Colleges in Relationship with Their Social, Economic, Political, and Educational Backgrounds and a Program for Their Improvement" (Ph.D. diss., Ohio State University, 1940), 75, 83.
4. Chapman, "Negro Land-Grant Colleges," 157. Opportunities for higher education for African Americans were subject to contraction, as well: six historically black schools offered medical education in 1900; by 1915, only Meharry and Howard offered medical and dental training. Those two schools trained 80 percent of black physicians and dentists until 1968. James E. Blackwell, *Mainstreaming Outsiders: The Production of Black Professionals* (Bayside, NY: General Hall, 1981), 13–14.
5. On the historic meaning of class, race, and gender identity in professional engineering, see David Noble, *America by Design: Science, Technology and the Rise of Corporate Capitalism* (Oxford: Oxford University Press, 1977); Amy Sue Bix, "Equipped for Life: Gendered Technical Training and Consumerism in Home Economics," *Technology and Culture* 43, no. 4 (2004): 728–754; Ruth Oldenziel, *Making Technology Masculine* (Amsterdam: Amsterdam University Press, 1999); Amy E. Slaton, *Reinforced Concrete and the Modernization of American Building, 1900–1930* (Baltimore: Johns Hopkins University Press, 2001); Jonson William Miller, "Citizen Soldiers and Professional Engineers: The Antebellum Engineering Culture of the Virginia Military Institute" (Ph.D. diss., Virginia Polytechnic and State University, 2008). On mid-century American understandings of aptitude and individual potential, see Michael Ackerman, "Mental Testing and the Expansion of Occupational Opportunity," *History of Education Quarterly* 35, no. 3: 279–300; for historical background on standardized testing and notions of workers' innate capacities, see Stephen Jay Gould, *The Mismeasure of Man*, revised and expanded edition (New York: W. W. Norton, 1996), 176–263.
6. On developments at Eastern Shore after Byrd's presidency, the most comprehensive source remains Carl S. Person, "Revitalization of an Historically Black College: A Maryland Eastern Shore Case" (Ph.D. diss., Virginia Polytechnic and State University, 1998). See also, Wilson H. Elkins, "A Decade of Progress

and Promise, 1954–1964: A Report to the Board of Regents of the University of Maryland" (Baltimore: University of Maryland, 1964); UMDES, "A Search for Truth through Quality Education" (Princess Anne, MD: University of Maryland, n.d.), both in UMDA; and Ruth H. Young, *Campus in Transition: University of Maryland Eastern Shore* (College Park, MD: University of Maryland, 1981). Material in the University of Maryland archives documents ongoing conflict on the campus under Byrd's successor, Wilson H. Elkins. See Wilson H. Elkins, "Statement in Response to Questions Raised by Students and Faculty Members Pertaining to Opportunities for Negroes in the University of Maryland," May 23, 1968, in UMDA.

7. UMDES, "University of Maryland, Eastern Shore, Princess Anne, Maryland: Report to the Middle States Association of Secondary Schools and Colleges," (1975), in UMDA; Person, "Revitalization of an Historically Black College," 43.

8. John M. Heffron, "Nation-Building for a Venerable South: Moral and Practical Uplift in the New Agricultural Education, 1900–1920," in *Essays in Twentieth-Century Southern Education: Exceptionalism and Its Limits*, ed. Wayne J. Urban (New York: Routledge, 1999), 58; Chapman, "Negro Land-Grant Colleges," 164. Similarly, Blackwell, *Mainstreaming Outsiders;* Frank Bowles and Frank A. deCosta, *Between Two Worlds: A Profile of Negro Higher Education* (New York: McGraw-Hill, 1971).

9. Chapman, "Negro Land-Grant Colleges," 81, 166–168; Linda R. Buchanan and Philo A. Hutcheson, "Reconsidering the Washingtion-Du Bois Debate: Two Black Colleges in 1910–1911," in *Essays in Twentieth-Century Southern Education: Exceptionalism and Its Limits*, ed. Wayne J. Urban (New York: Routledge, 1999); Heffron, "Nation-Building for a Venerable South"; William H. Watkins, *The White Architects of Black Education: Ideology and Power in America, 1865–1954* (New York: Teachers College Press, 2001).

10. Chapman, "Negro Land-Grant Colleges," 32, 65–72, 87–108; Maryland Commission on Higher Education of Negroes, "Report of the Maryland Commission on Higher Education of Negroes to the Governor and the General Assembly of Maryland" in *H.C. Byrd Papers* (UMDA), (1937); UMDES, "Report to the Middle States Association of Secondary Schools and Colleges," 3; Charles Rosenberg, "Science, Technology, and Economic Growth: The Case of the Agricultural Experiment Station Scientist, 1875–1914," in *No Other Gods: On Science and American Social Thought* (Baltimore: Johns Hopkins University Press, 1976 [1961]): 153–172.

11. John R. Thelin, *A History of American Higher Education* (Baltimore: Johns Hopkins University Press, 2004), 232; Chapman, "Negro Land-Grant Colleges," 78; Bowles and deCosta, *Between Two Worlds,* 3–4; UMDES, "Report to the Middle States Association of Secondary Schools and Colleges."

12. Young, *Campus in Transition: University of Maryland Eastern Shore,* 1–2; Person, "Revitalization of an Historically Black College," 44–48; Chapman, "Negro Land-Grant Colleges," 146.

13. Watkins, *The White Architects of Black Education,* 18–20; Thelin, *A History of American Higher Education,* 232.

14. Watkins, *The White Architects of Black Education*, 23. See also Nina Lerman, "From 'Useful Knowledge' To 'Habits of Industry': Gender, Race and Class in 19th-Century Technical Education" (Ph.D. diss., University of Pennsylvania, 1993).
15. Callcott, *A History of the University of Maryland*, 137–138; Heffron, "Nation-Building for a Venerable South." For a comprehensive discussion of southern class, race and gender ideologies in higher education in the 1800s, see Jonson William Miller, "Citizen Soldiers and Professional Engineers: The Antebellum Engineering Culture of the Virginia Military Institute" (Ph.D. diss., Virginia Polytechnic and State University, 2008).
16. Watkins, *The White Architects of Black Education*, 12–13. By 1930, over 65 percent of board members at a sampling of private and public universities and technical institutes in the United States were corporate lawyers, executives, and bankers. Thelin, *A History of American Higher Education*, 238.
17. Chapman, "Negro Land-Grant Colleges," 212.
18. Person, "Revitalization of an Historically Black College."
19. Author interview with George Callcott, December 1, 2001. Byrd may be seen as one of a number of "non-scholarly" university presidents in this period, including the head of Louisiana State University. Thelin, *A History of American Higher Education*, 246.
20. George H. Callcott, ed., *Forty Years a College President: Memoirs of Wilson Elkins* (Baltimore: University of Maryland, 1981); Andre Perry, "Where Is 'Our' Story within the Byrd Exhibit?" at www.studentorg.umd.edu (accessed 5/1/2007).
21. Thelin, *A History of American Higher Education*, 206–207.
22. Callcott, *A History of the University of Maryland*, 332.
23. George E. Gilbert and George E. Dieter, *Engineering at the University of Maryland College Park: The First 100 Years* (College Park, MD: University of Maryland, 1994), 16.
24. George W. Kable to H. C. Byrd, July 14, 1935, and Robert H. Spahr to H. C. Byrd, May 18, 1936, both in *H. C. Byrd Papers* (UMDA).
25. Karl Compton to H. C. Byrd, January 5, 1937; S. S. Steinberg to H. C. Byrd, February 8, 1937; H. C. Byrd to Karl Compton, April 24, 1937, all in *H. C. Byrd Papers* (UMDA); Gilbert and Dieter, *Engineering at the University of Maryland College Park*.
26. Gilbert and Dieter, *Engineering at the University of Maryland College Park*, 17; S. S. Steinberg to H. C. Byrd, February 8, 1937, in *H. C. Byrd Papers* (UMDA).
27. "Letter from H. C. Byrd to Editors," *Baltimore Sun*, November 10, 1937; Callcott, *A History of the University of Maryland*, 326–327.
28. "Letter from H. C. Byrd to Editors"; Maryland Commission on Higher Education of Negroes, "Report of the Maryland Commission on Higher Education of Negroes to the Governor and the General Assembly of Maryland," 81.
29. H. C. Byrd to Robert H. Spahr, July 18, 1936, in *H. C. Byrd Papers* (UMDA).

30. Maryland Commission on Higher Education of Negroes, "Report of the Maryland Commission on Higher Education of Negroes to the Governor and the General Assembly of Maryland," 12.
31. "Reprove a Man: A Colossal Liar," *Baltimore Afro-American*, May 25, 1935; Maryland Commission on Scholarships for Negroes, "Report of the Commission on Scholarships for Negroes to the Governor and Legislature of Maryland," (1939), in *H.C. Byrd Papers* (UMDA); Person, "Revitalization of an Historically Black College," 62. Byrd believed that others would go to jail with him in protest against the integration of the College Park campus; H. C. Byrd to Roger Howell, July 9 and 16, 1935, both in *H. C. Byrd Papers* (UMDA).
32. The state spending the second least was North Carolina, which still spent three times Maryland's investment; Maryland Commission on Higher Education of Negroes, "Report of the Maryland Commission on Higher Education of Negroes to the Governor and the General Assembly of Maryland," 8.
33. David Wharton, *A Struggle Worthy of Note: The Engineering and Technological Education of Black Americans* (Westport, CT: Greenwood Press, 1992), 68–69.
34. Chapman, "Negro Land-Grant Colleges," 217–219; Maryland Commission on Higher Education of Negroes, "Report of the Maryland Commission on Higher Education of Negroes to the Governor and the General Assembly of Maryland," 81.
35. Heffron, "Nation-Building for a Venerable South," 57–58; Slaton, *Reinforced Concrete and the Modernization of American Building, 1900–1930*, 50–61; Chapman, "Negro Land-Grant Colleges," 184.
36. On the "increasing concentration of Maryland's non-white population in the Baltimore area," see Eureal Grant Jackson, "Some Tendencies in Demographic Trends in Maryland, 1950–1956," *JNE* 26, no. 4 (1957): 515.
37. John Hardin Best, "Education in the Forming of the American South," in *Essays in Twentieth-Century Southern Education: Exceptionalism and Its Limits*, ed. Wayne J. Urban (New York: Routledge, 1999), 12; Milton H. Fies, "Research-Industry-and the South" (pamphlet produced by the Southern Association of Science and Industry, 1942), in *H.C. Byrd Papers* (UMDA).
38. Heffron, "Nation-Building for a Venerable South," 46–53. I am indebted to Elsa Barkley Brown for her comments on this subject.
39. William L. Fitzgerald, "A Suggested Program for the Betterment of the Negroes of Maryland," in *H. C. Byrd Papers* (UMDA), ca. 1935. Among those supporting such "gradualist" schemes were Anson Phelps Stokes Jr., scion of the family most involved with the philanthropic support of black education in this period through the Phelps-Stokes Fund. Stokes in turn supported the work of educational theorist Thomas Jesse Jones, identified by Watkins as the leading advocate of "safe, measured, gradual, orderly and minimal improvement" in black education. Watkins, *The White Architects of Black Education*, 95–96; also, Bowles and deCosta, *Between Two Worlds*.
40. Chapman, "Negro Land-Grant Colleges," 189, 245.
41. Best, "Education in the Forming of the American South"; Heffron, "Nation-Building for a Venerable South"; Thelin, *A History of American Higher Education*, 208.

42. Heffron, "Nation-Building for a Venerable South," 54–55, 67; Rosenberg, *No Other Gods;* Deborah Fitzgerald, *The Business of Breeding: Hybrid Corn in Illinois, 1890–1940* (Ithaca, NY: Cornell University Press, 1990); Chapman, "Negro Land-Grant Colleges," 84–106, 245.
43. William A. Link, "The School That Built a Town: Public Education and the Southern Social Landscape, 1880–1930," in *Essays in Twentieth-Century Southern Education*, ed. Wayne J. Urban (New York: Routledge, 1999), 28, 33.
44. William B. DeLauder, "The Hatch Act of 1887: Legacy, Challenges and Opportunities," speech delivered November 14, 2005 to the National Association of State Universities and Land Grant Colleges, at www.csrees.usda.gov (accessed 8/14/2007); "George Washington Carver Agricultural Experiment Station," at www.tuskegee.edu (accessed 6/21/2009).
45. Heffron, "Nation-Building for a Venerable South," 51; Maryland Commission on Higher Education of Negroes, "Report of the Maryland Commission on Higher Education of Negroes to the Governor and the General Assembly of Maryland"; Callcott, *A History of the University of Maryland*, 164–66; Chapman, "Negro Land-Grant Colleges," 74, 100.
46. Chapman, "Negro Land-Grant Colleges," 141–143.
47. Ibid., 116–118.
48. Ibid., 217.
49. Callcott, *A History of the University of Maryland*, 306; Blackwell, *Mainstreaming Outsiders*, 13–14.
50. Thurgood Marshall to H. C. Byrd, March 19, 1937, in *H. C. Byrd Papers*, UMDA.
51. Joseph F. Henry Jr., "Maryland's Colored Democracy," UMDA, ca. 1937.
52. Ambrose Caliver, *National Survey of the Higher Education of Negroes: A Summary* (Washington, DC: Federal Security Agency, U.S. Office of Education, 1943), 29–30; Blackwell, *Mainstreaming Outsiders*, 16.
53. H. C. Byrd to Luther N. Duncan, February 1, 1939, in *H. C. Byrd Papers*, UMDA.
54. J. W. Calhoun to H. C. Byrd, February 6, 1939, in *H. C. Byrd Papers*, UMDA.
55. Harmon Caldwell to H. C. Byrd, February 3, 1939, in *H. C. Byrd Papers*, UMDA.
56. Chapman, "Negro Land-Grant Colleges," 148.
57. Maryland Commission on Higher Education of Negroes, "Report of the Maryland Commission on Higher Education of Negroes to the Governor and the General Assembly of Maryland."
58. Chapman, "Negro Land-Grant Colleges," 252–258.
59. Ibid., 257.

3. The Disunity of Technical Knowledge

1. Ambrose Caliver, *National Survey of the Higher Education of Negroes: A Summary* (Washington, DC: Federal Security Agency, U.S. Office of Education, 1943), 44; The Office of Education reports on the "higher education of Negroes" through the war and postwar years aggregated "social, economic, and

educational data to indicate programs of higher education needed." The resulting survey was published in the form of multi-volume periodic reports and summaries. Anthony Welch, "Technocracy, Uncertainty, and Ethics: Comparative Education in an Era of Postmodernity and Globalization," in *Comparative Education: The Dialectic of the Global and the Local*, ed. Robert F. Arnove and Carlos Alberto Torres (Lanham, MD: Rowman & Littlefield, 2003), 26–27; Daniel J. Kevles, *The Physicists: The History of a Scientific Community in Modern America*, 2nd ed. (Cambridge: Harvard University Press, 1987), 295; Robert Kohler, Ellis Hawley, and Nathan Reingold, "Government Science," *Isis* 78, no. 4 (1987): 576–589; Thomas Borstelmann, *The Cold War and the Color Line* (Cambridge: Harvard University Press, 2001), 45–47. Compton quoted in Ambrose Caliver, *Education of Negro Leaders* (Washington, DC: Federal Security Agency, U.S. Office of Education, 1948), 57.

2. Sputnik, for example, which had actually demonstrated the potential of a non-capitalist state to achieve scientific mastery, somehow only bolstered U.S. technological triumphalism. One way or another, apparently, the West's egalitarian science would eventually overwhelm misguided Soviet applications of knowledge. See Joseph Farrell, "Equality of Education: A Half-Century of Comparative Evidence Seen from a New Millenium," in *Comparative Education*, ed. Robert F. Arnove and Carlos Alberto Torres, 6; Lorenzo Morris, *Elusive Equality: The Status of Black Americans in Higher Education* (Washington, DC: Howard University Press, 1979); Edward H. Berman, "Civic Education in the Corporatized Classroom," *Theory into Practice* 27, no. 4 (1988): 288.

3. Clifford T. McAvoy, quoted in "Laguardia Lauds N.Y.A. Training Plan," *New York Times*, December 19, 1942.

4. U.S. Department of War, "Utilization of Negro Manpower in the Postwar Army Policy," circular No. 124 (1946), 3–4, quoted in Caliver, *Education of Negro Leaders*, 48; Rufus B. Atwood, "The Origin and Development of the Negro Public College, with Especial Reference to the Land-Grant College," *JNE* 31, no. 3 (1962): 250; William Collins, "Race, Roosevelt and Wartime Production: Fair Employment in World War Two Labor Markets," *The American Economic Review* 91, no. 1 (2001): 272–286; Oscar James Chapman, "A Historical Study of Negro Land-Grant Colleges in Relationship with Their Social, Economic, Political, and Educational Backgrounds and a Program for Their Improvement" (Ph.D. diss., Ohio State University, 1940), 82, 345; Lee Finkle, "The Conservative Aims of Militant Leadership: Black Protest During World War Two," *Journal of American History* 60, no. 3 (1973): 703; Jennifer S. Light, *From Warfare to Welfare: Defense Intellectuals and Urban Problems in Cold War America* (Baltimore: Johns Hopkins University Press, 2003).

5. Philip A. Klinker, *The Unsteady March: The Rise and Decline of Racial Equality in America* (Chicago: University of Chicago Press, 1999), 148–49; Gladyce Helene Bradley, "Negro Higher and Professional Education in Maryland," *JNE* 17, no. 3 (1948): 374; Finkle, "The Conservative Aims of Militant Leadership"; UMD, Board of Regents, "Meeting" (1948), in UMDA, 5.

6. David Wharton, *A Struggle Worthy of Note: The Engineering and Technological Education of Black Americans* (Westport, CT: Greenwood Press, 1992);

Hilary Herbold, "Never a Level Playing Field: Blacks and the GI Bill," *JBHE*, no. 6 (1994/1995): 104–108; Roger L. Geiger, "Milking the Sacred Cow: Research and the Quest for Useful Knowledge in the American University since 1920," *Science, Technology & Human Values* 13, no. 3/4 (1988): 334.

7. H. C. Byrd to R. A. Grigsby, May 21, 1947, in *H. C. Byrd Papers*, UMDA; Maryland State College, "Maryland State College Bulletin," UMDA, ca. 1950.
8. Citizens Committee on Higher Education, "Some Facts About the Higher Education of Colored People in Maryland " (pamphlet, 1949), in *H. C. Byrd Papers*, UMDA, 9.
9. "A Plain Case of Waste," *Baltimore Afro-American*, December 7, 1946.
10. Ibid.
11. H. C. Byrd to H. T. Heald, October 15, 1942, in *H. C. Byrd Papers*, UMDA; George H. Callcott, *A History of the University of Maryland* (Baltimore: Maryland Historical Society, 1966), 3–4.
12. Elder H. Russell to T. Walter Gough, June 13, 1941, and J. W. Jewett to H. C. Byrd, April 1, 1941, both in *H. C. Byrd Papers*, UMDA.
13. "Judge Soper Urges Further Expansion of Morgan College," *Baltimore Sun*, December 7, 1949; Finkle, "The Conservative Aims of Militant Leadership," 699.
14. Caliver, *National Survey*.
15. Ibid., 43. Howard University's wartime experiences may be the exception that proves the rule. Howard physicist Herman Branson directed research on AEC projects there during the war, and reported an "extremely low level of prejudice among scientists" engaged in war research. "Physicists of the African Diaspora: Herman Branson" at www.math.buffalo.edu (accessed 6/15/2006); Caliver, *Education of Negro Leaders*, 43.
16. Commission on Problems Affecting Negro Population, "Report of the Governor's Commission on Problems Affecting the Negro Population (Maryland)," (1943), 75, in UMDA; Klinker, *The Unsteady March*, 149.
17. Geiger, "Milking the Sacred Cow," 335.
18. "Memorandum (on the Establishment of Engineering Facilities at College Park)," in *H. C. Byrd Papers*, UMDA, 5.
19. Ibid.
20. Carl S. Person, "Revitalization of an Historically Black College: A Maryland Eastern Shore Case" (Ph.D. diss., Virginia Polytechnic and State University, 1998).
21. Thomas O'Neill, "Princess Anne Referendum Move Made," *Baltimore Sun*, April 7, 1949; Person, "Revitalization of an Historically Black College," 46, 64–65; William L. Marbury, *Higher Education in Maryland* (Washington, DC: American Council on Education, 1947).
22. Morris A. Soper, "Memorandum of Remarks by Judge Soper before the Legislative Council on Recommendations of the Maryland Commission on Higher Education" (1947), in *H. C. Byrd Papers*, UMDA. Soper's outlook on race relations seems to have been progressive relative to Byrd's, but in some ways mixed. In 1949, aware of black unrest regarding segregation in Maryland, Soper referred to blacks in Maryland as "on the march" and warned white

Marylanders to recall the "savage reaction of the suppressed Japanese in the last war." Maryland Commission on Higher Education of Negroes, "Minutes, State Commission on Negro Higher Education Tuesday, December 6, 1949," in UMDA, 1.

23. "Brief Statement for the Committee on Education of the General Assembly About Higher Education for Negroes with Recommendations" (typescript), in *H. C. Byrd Papers*, UMDA (1951), 19; R. M. Stewart, "Tentative Suggestions for Consideration by Princess Anne College," UMDA (1948), 14; Maryland State College, "Bulletin;" Carl S. Person, "Revitalization of an Historically Black College," 60–64.

24. Perhaps most damningly, of $6.3 million received by the UMD system from the federal government between 1938 (when it purchased Eastern Shore from Morgan State College) and 1947, only $153,241 of that money had been spent at Eastern Shore. Citizens Committee on Higher Education, "Some Facts about the Higher Education," 4. Byrd characterized materials issued by the organization as "a lot of plain, unadulterated misstatements"; H. C. Byrd to Glenn Brown, January 25, 1949, in *H. C. Byrd Papers*, UMDA.

25. O'Neill, "Princess Anne Referendum Move Made."

26. Citizens Committee on Higher Education, "Some Facts about the Higher Education"; Citizens Committee on Higher Education, "Agenda" (1949), in *H. C. Byrd Papers*, UMDA; "Brief Statement for the Committee on Education [1951]," 5. Johns Hopkins had been using some state funds for educating engineers, but had provided many facilities out of its private resources; "College Heads Cautious on School Report," *Baltimore Sun*, February 3, 1947.

27. George Streator, "Negroes Criticize Maryland School," *New York Times*, December 5, 1948; Person, "Revitalization of an Historically Black College."

28. Gregory Kannerstein, "Black Colleges: Self-Concept," in *Black Colleges in America*, ed. Charles V. Willie and Ronald R. Edmonds (New York: Teachers College Press, 1978); John R. Hill, "Presidential Perception: Administrative Problems and the Needs of Public Black Colleges," *JNE* 44, no. 1 (1975): 53–62; Gerald L. Smith, *A Black Educator in the Segregated South: Kentucky's Rufus B. Atwood* (Lexington: University Press of Kentucky, 1994); Wayne J. Urban, "Book Review," *American Historical Review* 100, no. 5 (1995): 1721–1722. Urban concisely notes that "[t]he often, although not always, accurate image of the black college president as tyrant demands attention" and that "the relations between civil rights activist students in the 1950s and 1960s and a black college president constitute a troubling episode for any presidential biographer" (p. 1721).

29. Williams's departure from Eastern Shore after 23 years as president was accompanied by a fair amount of ill feeling; see John T. Williams to Blair Lee III, May 26, 1970, in *UMDES Records*, UMDA; "Maryland State College," *Philadelphia Inquirer*, August 17, 1948; Black Coalition of the University of Maryland, "An Investigative Study of Maryland State College" (1970), in UMDA; "Williams, Retired President of Shore College, Dies at 66," *Evening Sun*, July 14, 1971; Mary Corddry, "Shore Educator Blasts His Foes," *Baltimore Sun*, August 16,

1970. Person describes Williams's shifting relationship with the white and black communities of Princess Anne and the college itself in "Revitalization of an Historically Black College."
30. John T. Williams, "Typescript of Testimony before the Legislative Council of the Maryland Legislature" (1948), in UMDA; Soper, "Memorandum of Remarks by Judge," 6.
31. Williams, "Typescript of Testimony," 4. Byrd frequently defended conditions at Eastern Shore with the argument that the state had failed to appropriate sufficient funds for the upkeep of that campus; see H. C. Byrd to Evelyn Andrews, October 24, 1949, "Legislative Testimony: Princess Anne College" (typescript), in *H. C. Byrd Papers*, UMDA.
32. Williams, "Typescript of Testimony," 6. Similarly, in legislative testimony a few days later, likely composed by Byrd, "The college at Princess Anne has been recognized by the white people of the Eastern Shore as a splendid influence in maintaining good relationships between the two races." "Legislative Testimony: Princess Anne College," 8.
33. Williams, "Typescript of Testimony."
34. H. C. Byrd to Milton Brown, June 10, 1948.
35. "Legislative Testimony: Princess Anne College," 8–9.
36. John T. Williams, "The Challenge" (typescript, 1948), in *H. C. Byrd Papers*, UMDA, 11.
37. "Elkins Urged to Intervene in Princess Anne Dispute," *Baltimore Sun*, March 4, 1964; "Dogs and Autos Break up March," *Baltimore Sun*, February 24, 1964; Charles Rabb, "Restaurant Owners Hold Key to Princess Anne Flare-up," *Washington Post*, February 26, 1964.
38. Cited in Citizens Committee on Higher Education, "Some Facts about the Higher Education," 4.
39. Stewart, "Tentative Suggestions for Consideration by Princess Anne College." Groups opposed to moving federal land-grant funds from the campus in Princess Anne to Morgan State College could readily call on arguments that Baltimore was no place for a college intended to offer agricultural instruction. President Williams, defending the build-up of Eastern Shore in legislative testimony in the late 1940s, supported that argument with somewhat hollow logic. He made the point that land-grant colleges should be rural. He invoked as examples Penn State and Cornell, and even Prairie View A&M University, the black branch of the Texas A&M land-grant system. Prairie View, Williams noted, was "forty miles from nowhere" but apparently still served its constituency well. He did not mention that that school, like Eastern Shore, was perpetually neglected by its state sponsors (see Chapters 5 and 6). Williams, "Typescript of Testimony before the Legislative Council of the Maryland Legislature."
40. Stewart, "Tentative Suggestions for Consideration by Princess Anne College," 2–3.
41. Ibid., 2.
42. H. C. Byrd to R. M. Stewart, June 9, 1948, in "Legislative Testimony: Princess Anne College" (typescript), in *H. C. Byrd Papers*, UMDA.
43. H. C. Byrd to Fred J. Kelly, April 4, 1949, in *H. C. Byrd Papers*, UMDA.

44. "Legislative Testimony: Princess Anne College," (typescript), 8.
45. Caliver, *Education of Negro Leaders*, 2. The report later offers a brief disavowal of this point: "It should be emphasized that there is no assumption here that Negroes and whites must necessarily serve only members of their respective groups"; then adds, "But the more important fact is that there are not enough quality professionally trained persons to render the needed services of *either group* [emphasis added]." The proposition that the citizenry can be divided into "groups" disaggregates portions of the population on the basis of race (p. 52).
46. Ibid., 44.
47. Peyton S. Hutchison, "Marginal Man with a Marginal Mission: A Study of the Administrative Strategies of Ambrose Caliver, Black Administrator of Negro and Adult Education, the United States Office of Education, 1930–1962" (Ph.D. diss., Michigan State University, 1975). On the role of higher education in workforce stratification and control, see David K. Brown, "The Social Sources of Educational Credentialism: Status Cultures, Labor Markets, and Organizations," *Sociology of Education* 74 (2001): 19–34. On engineering as a case of civic enrollment, see David Noble, *America by Design: Science, Technology and the Rise of Corporate Capitalism* (Oxford: Oxford University Press, 1977).
48. Dan S. Green and Edwin D. Driver, "Introduction," in W. E. B. Du Bois, Dan S. Green, and Edwin D. Driver, *W. E. B. Du Bois: On Sociology and the Black Community* (Chicago: University of Chicago Press, 1978): 1–51.
49. Caliver, *National Survey*.
50. Ibid., 9; Stewart, "Tentative Suggestions for Consideration by Princess Anne College," 11.
51. "Legislative Testimony: Princess Anne College" (typescript), 8.
52. Caliver, *Education of Negro Leaders*, 16.
53. The *Philadelphia Inquirer*, entirely laudatory regarding programming at Eastern Shore, described the college as serving to "motivate the individual to become a more effective worker" and "good citizen"; "Maryland State College," *Philadelphia Inquirer*, August 17, 1948.
54. Ironically, some of the most influential African Americans disdained these prescriptions. Labor leader A. Philip Randolph, disturbed by black distaste for union activism in the 1940s, tellingly used the phrase, "the leadership will have to catch up with the followship." Quoted in Finkle, "The Conservative Aims of Militant Leadership," 697.
55. Caliver, *Education of Negro Leaders*, 17.
56. Ibid., 47; U.S. Department of War, Bureau of Public Relations, "Report of Board of Officers on Utilization of Negro Manpower in the Post-War Army" (press release, 1946).
57. Caliver, *National Survey*, 19, 41.
58. Caliver, *Education of Negro Leaders*, 13–14; Hutchison, "Marginal Man," 147–148. On colonial perceptions of skill and aptitude among indigenous peoples, see Warwick Anderson, "Introduction: Postcolonial Technoscience," *Social Studies of Science* 32, no. 5/6 (2002): 643–658; Karen Kupperman, *Settling with the Indians* (Totowa, NJ: Rowman & Littlefield, 1980); Patrick

Malone, *The Skulking Way of War: Technology and Tactics among the New England Indians* (Baltimore: Johns Hopkins University Press, 1993).

59. Caliver, *Education of Negro Leaders*, 12; Caliver, *National Survey*, 31.
60. Caliver, *Education of Negro Leaders*, 12–13.
61. Kannerstein, "Black Colleges: Self-Concept"; Person, "Revitalization of an Historically Black College." See also American Association of University Professors Committee on Historically Black Institutions and Scholars of Color, "Historically Black Colleges and Universities: Recent Trends (2007)" (2006), at www.aaup.org (accessed 8/12/2008).
62. Caliver, *National Survey*, 27, 44–45; Caliver, *Education of Negro Leaders*, 13–14.
63. Caliver, *Education of Negro Leaders*, 16, 22, 23. Similarly, the Office of Education's 1943 report ascribed poor diet and other habits of African Americans as limiting their ability to perform in challenging educational and professional settings. Caliver, *National Survey*, 40. Many critics recognized the arbitrary nature of such causal explanations for low black participation in higher education. Howard H. Murphy, Chairman of the Maryland Committee for Equal Educational Opportunities, explained to UMD's Board of Regents in 1948 that many students who were interested in higher education failed to apply, fearing rejection. UMD, Board of Regents, "Meeting [1948]," 14.
64. Caliver, *Education of Negro Leaders*, 16.
65. Caliver, *National Survey*, 43.
66. Ibid., 44.
67. Ibid., 47.
68. UMD, Board of Regents, "Meeting [1948]," 7–8.
69. "Md. U. Seeks Legislation on Negro Entry," *Baltimore Times Herald*, November 19, 1948.
70. "U. Of M. Must Admit Negroes, Byrd Says," *Washington Post*, November 15, 1948; University of Maryland, Board of Regents, "Meeting [1948]." A year later, Byrd declared, "It is not to be questioned that, under present State policies, the college will be developed to have equal facilities and just as high standards as the Land Grant College of the University at College Park for white people." H. C. Byrd to John T. Williams, February 19, 1949, in *H. C. Byrd Papers*, UMDA. On Byrd's inconsistency in this period, see also "Judge Soper Urges Further Expansion of Morgan College."
71. James A. Dombrowski to H. C. Byrd, August 10, 1949, and H. C. Byrd to James A. Dombrowski, August 22, 1949, in *H. C. Byrd Papers*, UMDA.
72. "Two Students, Poor Library at Teachers' Summer School," *Baltimore Afro-American*, June 22, 1948; "2 Out-of-Town Students Defy Community Feeling" and Editorial, "Byrd Should Cut this Foolishness," *Baltimore Afro-American*, June 26, 1948.
73. "U. of M. Takes No Action on Negro Ruling," *Baltimore Sun*, April 16, 1950; "Brief Statement for the Committee [1951]"; Rita Sutter, "When Yesterday's Traditions Are Thankfully Past," *Outlook* 8, no. 15 (1994); UMD, Board of Regents, "Meeting [1948]," and UMD, Board of Regents, "Meeting [1950]" UMDA.

74. H. C. Byrd to Dean H. Boyd Wylie, April 25, 1951, in *H. C. Byrd Papers*, UMDA.
75. "Byrd Upholds University's Nursing School," *Baltimore Sun,* October 14, 1949; "Brief Statement for the Committee [1951]," 6.
76. University of Maryland Board of Regents, "Recent Decisions" (typescript, ca. 1952), in *H. C. Byrd Papers,* UMDA.
77. Sutter, "When Yesterday's Traditions Are Thankfully Past," 6.
78. "Brief Statement for the Committee [1951]," 6. This statement, entirely hostile to the conduct of Morgan State College and its supporters, is not attributed, but the language is nearly identical to that which Byrd uses elsewhere: "A good deal more might be written as to the history of commissions that have studied higher education for Negroes, but no particularly good ends would be served in dealing with personalities and questionable ethical practices" (pp. 21–22). At one point, the author writes, "However, the President of the University of Maryland must observe that he did not expect this criticism to result in attacks on the University of Maryland" (p. 20).
79. Ibid., 7.
80. Ibid., 14. H. C. Byrd to John T. Williams, February 19, 1949, in *H. C. Byrd Papers,* UMDA.
81. "Brief Statement for the Committee [1951]," 8.
82. UMD, Board of Regents, "Recent Decisions," 5.
83. Callcott, *A History of the University of Maryland,* 353; Arthur T. Johnson and Rosemary L. Parker, "The Effects of Technology on Diversity or, When Is Diversity Not Diversity?" (Paper presented at the American Society for Engineering Education Annual Conference and Exposition, 2001); "The Sharp Dropoff in Black Engineering Students," *JBHE* 21 (1998): 31–32; Person, "Revitalization of an Historically Black College"; Ruth H. Young, *Campus in Transition: University of Maryland Eastern Shore* (College Park, MD: University of Maryland, 1981); UMDES, "A Search for Truth through Quality Education" (Princess Anne, MD: University of Maryland, n.d.). An index of UMD's more recent commitment to black participation in engineering can be seen in 1998 statistics regarding black enrollment at twenty-five of the nation's top-rated graduate schools of engineering; at 5.8 percent, UMD ranked second, exceeded only by Georgia Tech (6.5 percent). Tellingly, of the twenty-five graduate schools, twenty-two showed black enrollment below 4 percent ("The Sharp Dropoff in Black Engineering Students," 32).

4. Opportunity in the City

1. On contemporary ideologies surrounding education and equity, see Michael F. D. Young, ed., *Knowledge and Control: New Directions for the Sociology of Education* (London: Collier-Macmillan, 1971); and Michael Ackerman, "Mental Testing and the Expansion of Occupational Opportunity," *History of Education Quarterly* 35, no. 3: 279–300.
2. Robert M. Fogelson, Gordon S. Black, and Michael Lipsky, "Review Symposium," *American Political Science Review* 63, no. 4 (1969): 1269–1281.

3. Chicago Department of Development and Planning, "The Comprehensive Plan of Chicago, Conditions and Trends: Population, Economy, Land" (Staff Study Report, 1967), in CHM; Roger Biles, *Richard J. Daley: Politics, Race and the Governing of Chicago* (Dekalb, IL: Northern Illinois University, 1995); Arnold R. Hirsch, *Making the Second Ghetto: Race and Housing in Chicago 1940–1960* (Chicago: University of Chicago Press, 2nd ed., 1998); Urban Associates of Chicago, "A Report to the Department of Urban Renewal/City of Chicago on an Evaluation of Industrial Centers in Urban Renewal Areas" (Chicago: 1975), in CHM; University of Illinois Senate-UI CUD Committee on Curricular Expansion, "A Concept for the University of Illinois in Chicago Lincoln Campus," 1959, UICA.
4. Hirsch, *Making the Second Ghetto*.
5. Chicago Riot Study Committee, "Report of the Chicago Riot Study Committee to the Hon. Richard J. Daley" (Chicago: 1968), 31. The Riot Study Committee surveyed city efforts to alleviate poverty, and mentioned postsecondary education only in the context of a single Upward Bound initiative (p. 145).
6. William DeFotis, *A History of the College of Engineering at the University of Illinois (1946–2001)* (Chicago: University of Illinois at Chicago, 2001); Randall Collins, "Functional and Conflict Theories of Educational Stratification," *American Sociological Review* 36 (1971): 1002–1019. See also Michael N. Bastedo and Patricia J. Gumport, "Access to What? Mission Differentiation and Academic Stratification in U.S. Public Higher Education," *Higher Education* 46, no. 3 (2003): 341–359.
7. Gary S. May and Daryl E. Chubin, "A Retrospective on Undergraduate Engineering Success for Underrepresented Minority Students," *Journal of Engineering Education* 92, no. 1 (2003): 1–13; Ruth Oldenziel, *Making Technology Masculine* (Amsterdam: Amsterdam University Press, 1999).
8. Michael Lipsky, in Fogelson, Black, and Lipsky, "Review Symposium," 1281.
9. Raymond Murphy, "Power and Autonomy in the Sociology of Education," *Theory and Society* 11, no. 2 (1982): 196; Collins, "Functional and Conflict Theories," 1007.
10. Hirsch, *Making the Second Ghetto*, 171–180.
11. Armour Institute of Technology, "A New Home for Chicago's Center of Education in Engineering and Architecture" (1935) in IITA; Robert H. Kargon and Scott G. Knowles, "Knowledge for Use: Science, Higher Learning, and America's New Industrial Heartland, 1880–1915," *Annals of Science* 59 (2002): 1–20; Kevin Harrington, "Hilbersheimer and the Redevelopment of the South Side of Chicago," in *In the Shadow of Mies: Ludwig Hilbersheimer, Architect, Educator and Urban Planner*, ed. Richard Pommer, David Spaeth, and Kevin Harrington (Chicago: Art Institute of Chicago/Rizzoli, 1988).
12. IIT, "Illinois Institute of Technology" (brochure, 1950), in IITA; Hirsch, *Making the Second Ghetto*; Daniel Bluestone, "Chicago's Mecca Flat Blues," *Journal of the Society of Architectural Historians* 57, no. 4 (1998): 382–403.
13. Harrington, "Hilbersheimer and the Redevelopment of the South Side of Chicago," 76; Hirsch, *Making the Second Ghetto*.
14. IIT, "IIT Hall of Fame: Henry Heald," at www.IIT.edu (accessed 1/12/2006).

15. "Chicago's Blight War: A Tremendous Effort," *Defender*, January 9, 1960; Hirsch, *Making the Second Ghetto*, 103–114.
16. Bluestone, "Chicago's Mecca Flat Blues," 394–401. Ninety percent of the low-income units removed for urban renewal were never replaced. New commercial, industrial, and municipal projects occupied more than 80 percent of the land cleared for these projects. See George Lipsitz, "The Possessive Investment in Whiteness: Racialized Social Democracy and The 'White' Problem in American Studies," *American Quarterly* 47, no. 3 (1995): 369–387. What is more, as Hirsch indicates, when private interests oversaw the provision of low-income housing in Chicago they commonly reiterated existing geographic racial divisions. Hirsch, *Making the Second Ghetto*, 263.
17. "Illinois Institute of Technology" (brochure, 1950) in IITA; Bluestone, "Chicago's Mecca Flat Blues," 395–396; Hirsch, *Making the Second Ghetto*.
18. Mary Corbin Sies and Christopher Silver, eds., *Planning the Twentieth-Century American City* (Baltimore: Johns Hopkins University Press, 1996); June Manning Thomas and Marsha Ritzdorf, eds., *Urban Planning and the African American Community: In the Shadows* (Thousand Oaks, CA: Sage Publications, 1997).
19. "Special Report: Illinois Institute of Technology," *Plans and Progress: Chicago Plan Commission*, December 1954, in IITA; Harrington, "Hilbersheimer and the Redevelopment of the South Side of Chicago"; Bluestone, "Chicago's Mecca Flat Blues"; Taleb Dia Ed-deen Ar-Ritai, "The New University Environment: A Twentieth-Century Urban Ideal" (Ph.D. diss., University of Pennsylvania, 1983), 84–88.
20. Editorial, "A Great School Plans Expansion," *Chicago Tribune*, May 28, 1965. On connections between utilitarian building design and cultural ideologies regarding technical expertise, see Amy E. Slaton, *Reinforced Concrete and the Modernization of American Building, 1900–1930* (Baltimore: Johns Hopkins University Press, 2001). This set of aesthetic choices has been ably analyzed by Kevin Harrington, who identifies a racial element to the social conservatism scholars now attribute to many examples of high-modernist building design. Mies's futuristic designs for the new IIT buildings, and those of his colleagues who helped plan and landscape the campus, did not openly address the changing social relations enacted by IIT's expansion, at best presuming an automatic improvement in the community and at worst passively consenting to the displacement of poor, nonwhite residents. See Harrington, "Hilbersheimer and the Redevelopment of the South Side of Chicago."
21. IIT, "Living at IIT" (pamphlet, n.d.), in IITA; Bluestone, "Chicago's Mecca Flat Blues."
22. Hugh Folk, *The Shortage of Scientists and Engineers* (Lexington, MA: Heath Lexington Books, 1970).
23. John T. Rettaliata, "The Financial Future of Private Higher Education in the United States" (1969), in IITA, 14.
24. IIT, "The University of Science and Technology: A Statement of Purpose" (report, 1966), in IITA; "Research for Industry: Armour Research Foundation of Illinois Institute of Technology," *Cenco News Chats/Central Scientific Com-*

pany (November, 1954), in IITA; Bluestone, "Chicago's Mecca Flat Blues," 397.
25. "The University of Science and Technology: A Statement of Purpose." Prevailing definitions of the city's boundaries could justify the institution's sense of its own mission in regard to education, research, or other economic contributions; see George Bugliarello and Harold A. Simon, "The Impact of the University on Its Environment: The Regional Role of Engineering Colleges," (College of Engineering, University of Illinois Chicago, 1973), in UICA, 12.
26. John T. Rettaliata, "Press Release" (1952), in IITA.
27. "Illinois Institute of Technology" (brochure, 1950), in IITA, 3.
28. Chicago Riot Study Committee, "Report to the Hon. Richard J. Daley," 34.
29. Ibid., 64.
30. Gary Marx, "Two Cheers for the National Riot (Kerner) Commission Report," in *Black Americans: A Second Look*, ed. J. F. Szwed (New York: Basic Books, 1970), at http://web.mit.edu (accessed on 8/14/2008).
31. Chicago Riot Study Committee, "Report to the Hon. Richard J. Daley," 31.
32. IIT, "One-hundred children of the Angel Guardian orphanage . . ." [press release, June 19, 1951], in IITA; IIT, "A bus load of children from the Ada S. McKinley community house . . ." [press release, May 9, 1951]; William R. Hammond, Ada S. McKinley Community House, "Staff members of Illinois Institute of Technology and Armour Research Foundation contribute $1,680 . . ." [press release, April 25, 1952]; "IIT Woman's Club Offers Active Program," *IIT Reports* 7, no. 9 (September-October, 1968), 7–8; all in IITA; IIT, "President's Report 1966–1967," in *Papers of John Rettaliata* (IITA).
33. University of Illinois Chicago, "UICC 1979 Mission Statement" (1980), in UICA.
34. George Rosen, *Decision Making Chicago-Style: The Genesis of a University of Illinois Campus* (Urbana, IL: University of Illinois, 1980); University of Illinois Chicago, "University of Illinois at Chicago Circle: Location" (1964), in UICA; Biles, *Richard J. Daley*.
35. Mike Grunsten and Jim Schaefer, "An Interview with the Dean of Engineering," *Chicago Circle Engineering* 2, no. 1 (1970).
36. All campus buildings were concentrated on a third of this acreage, to achieve a compact, "city like" facility. See Taleb Dia Ed-Deen Ar-Rifai, "The New University Environment: A 20th Century Urban Ideal" (Ph.D. diss., University of Pennsylvania, 1983), 97–110; "Campus City, Continued," *Architectural Forum* (1968), 27.
37. University of Illinois Public Information Office, "Statement on Minority Employment" (1969), in UICA; ABLA Community Center Local Advisory Council, "A Development Plan for Establishing and Operating the ABLA Community Center," in UICA; Dennis Petrick, "Urban Renewal Aims to Fortify Economic Status of CC Area," *Chicago Illini*, February 1, 1965.
38. Biles, *Richard J. Daley*.
39. Alan B. Anderson and George W. Pickering, *Confronting the Color Line: The Broken Promise of the Civil Rights Movement in Chicago* (Athens, GA: University of Georgia Press, 1986); Hirsch, *Making the Second Ghetto;* Robert A.

Beauregard and Briavel Holcomb, "Dominant Enterprises and Acquiescent Communities: The Private Sector and Urban Revitalization," *Urbanism Past and Present* 8 (1979): 18–31.

40. Yale Rabin, "The Persistence of Racial Isolation: The Role of Government Action or Inaction," in *Urban Planning and the African American Community*, ed. June Manning Thomas and Marsha Ritzdorf (Thousand Oaks, CA: Sage Publications, 1997), 95–98.

41. John T. Rettaliata, "Remarks by Dr. John Rettaliata at the Alumni Reception" (1969), in IITA, 1.

42. Kenneth A. McCollom, "Professional Schools of Engineering," *Engineering Education* (1972): 915–918; Oldenziel, *Making Technology Masculine*, 53–54; Steven B. Sample, "The Evolution of Engineering Education," *Chicago Circle Engineering* 6, no. 3 (1972).

43. Bugliarello and Simon, "The Impact of the University," 49.

44. Harold S. McFarland, "Minority Group Employment at General Motors," 131–136, and Gerry E. Morse, "Equal Employment Opportunity at Honeywell, Inc.," 123–124, in *The Negro and Employment Opportunity: Problems and Practices*, ed. Herbert R. Northrup and Richard L. Rowan (Ann Arbor, MI: University of Michigan, 1965).

45. "Chicago's Blight War"; Anderson and Pickering, *Confronting the Color Line*, 168–171; Herbert R. Northrup and Richard L. Rowan, eds., *The Negro and Employment Opportunities: Problems and Practices*; Maureen T. Hallinan, "Sociological Perspectives on Black-White Inequalities in American Schooling," *Sociology of Education* 74, extra issue (2001); Biles, *Richard J. Daley*, 103–118.

46. Marx, "Two Cheers for the National Riot (Kerner) Commission Report"; Claire Jean Kim, "Clinton's Race Initiative: Recasting the American Dilemma," *Polity* 33, no. 2 (2000), 183.

47. Jiannbin Lee Shiao, "Beyond States and Movements: Philanthropic Contributions to Racial Formation in Two U.S. Cities, 1960–1990" (Ph.D. diss., University of California, 1998); John U. Monro, "College Programs for Disadvantaged Youth," *Expanding Opportunities: The Negro and Higher Education* 1, no. 3 (1964).

48. D. E. Irwin, "Discussion" (Paper presented at the American Society for Engineering Education 25th Annual College-Industry Conference, Stanford University, 1973).

49. "A Push for Black Engineers," *Business Week* (February 14, 1977): 124.

50. IIT, Board of Trustees, "Minutes" (1974), in IITA; IIT, Board of Trustees, "Minutes" (1975), in IITA, 19.

51. "54 Juniors in IIT Program," *Chicago Tribune*, July 18, 1974; "New Committee Chairman," *Minorities in Engineering (Assembly of Engineering/National Research Council)* 3, no. 2 (1977); IIT, "President's Report" (1973–1974), in IITA, 3–4.

52. Hallinan, "Sociological Perspectives on Black-White Inequalities in American Schooling"; Carolyn Cummings Perrucci and Robert Perrucci, "Social Origins,

Educational Contexts, Career Mobility," *American Sociological Review* 35, no. 3 (1970): 451–463.
53. IIT, "Illinois Institute of Technology Minorities in Engineering Program: 1980 Report" (1980), in IITA, 2.
54. Thomas L. Martin, Jr., untitled manuscript (May 20, 1976), in IITA, 5.
55. IIT, "Minorities in Engineering Program: First Annual Report" (September 1, 1975), in IITA, 1.
56. Thomas L. Martin, Jr., speech to RCA (June 13, 1978), in IITA; Institutional Study Committee, "Report of the Institutional Study Committee of Illinois Institute of Technology on Goals of Engineering Education/American Society for Engineering Education" (report, 1964), in IITA; Michael F. D. Young, "An Approach to the Study of Curricula as Socially Organized Knowledge," in *Knowledge and Control*, 37; Thomas L. Martin Jr., "The Minorities in Engineering Program" (1978), in IITA; Peter Chiarulli, "Draft No. 1, April 16, 1964: Present and Future Goals of Engineering Education" (1964), in IITA.
57. National Academy of Sciences Committee on Minorities in Engineering, "Building the Multiplier Effect: Summary of a National Symposium September 14–16, 1978"; Donald G. Dickason, "An Assessment of Elements Present in Successful Minority Student Programs in Colleges of Engineering," 39; National Academy of Engineering Commission on Education, "The National Academy of Engineering Role in a Program to Increase the Number of Minorities in Engineering," 9; all in *Central File (Assembly)*, National Academy of Engineering Archives, Washington, DC. Also, IIT, "Minorities in Engineering Program: 1980 Report," 1; Elizabeth A. Duffy and Idana Goldberg, *Crafting a Class: College Admissions and Financial Aid, 1955–1994* (Princeton, NJ: Princeton University Press, 1998), 152, cited in W.G Bowen and D. Bok, *The Shape of the River: Long-Term Consequences of Considering Race in College and University Admissions* (Princeton, NJ: Princeton University Press, 1998), 8.
58. Morse, "Equal Employment Opportunity at Honeywell, Inc.," 124. On hesitations within high-tech industry regarding the competence of minority personnel in this period, see Terry H. Anderson, *The Pursuit of Fairness: A History of Affirmative Action* (Oxford, Oxford University Press, 2004), 62-64.
59. Young, "An Approach to the Study of Curricula as Socially Organized Knowledge," 25.
60. Sociologists of education articulated the idea that failure helped define success: "In so far as [weaker] pupils' behavior is explicitly seen by teachers as inappropriate or inadequate, it makes more visible or available what is held to be appropriate pupil behavior," in Nell Keddie, "Classroom Knowledge," in *Knowledge and Control*, 134.
61. Martin, "The Minorities in Engineering Program."
62. Thomas L. Martin, Jr., untitled manuscript (May 20, 1976), in IITA, 4.
63. Dickason, "An Assessment of Elements Present in Successful Minority Student Programs in Colleges of Engineering," 49; "Enrollment and Student Statistics for the 1980–81 Year," *IIT Chronicle* 2, no. 3 (1981); Table 15, "From E. F.

Steuben, Director, Center for Educational Development to IIT Administrators" (1980), in IITA.

64. "For Minorities . . ." *Defender,* March 25, 1974; "From J. G. Mikota to H. L. Phillips, Jr., Continental Can Company" (1974), in IITA.

65. NACME, "Nearly 30 Years of Leadership and Support" at www.nacme.org (accessed 6/6/2006); "A Faltering Attempt to Train Minority Engineers," *Business Week* (July 2, 1979): 78J.

66. NACME, "Nearly 30 Years of Leadership and Support."

67. Many studies of this era indicated that factors such as father's occupation or family income level determined a student's likelihood of attending college. One large 1970 study found that college populations were strongly skewed toward the wealthiest sectors of American society; Folk, *The Shortage of Scientists and Engineers,* 14–15. Similarly, Patricia Gurin and Edgar Epps, *Black Consciousness, Identity, and Achievement: A Study of Students in Historically Black Colleges* (New York: Wiley, 1975), 195, 329.

68. Gurin and Epps, *Black Consciousness, Identity, and Achievement.*

69. Kevin Oliver and Michael Hannafin, "Developing and Refining Mental Models in Open-Ended Learning Environments: A Case Study," *Educational Technology, Research and Development* 49, no. 4 (2001): 5–33; Elaine Seymour and Nancy Hewitt, *Talking About Leaving* (Boulder, CO: Westview Press, 1997). On the attribution of high cultural status to knowledge acquired through avoidance of "group work or co-operativeness," see Young, "An Approach to the Study of Curricula as Socially Organized Knowledge," 38. Basil Bernstein, describing education in the United States and England in the 1960s, wrote that the "pacing of knowledge (i.e. the rate of expected learning) is implicitly based upon the middle-class socialization of the child," a socialization which "immensely facilitates" school learning. He labeled this kind of background a "hidden subsidy" for middle-class students. Basil Bernstein, "On the Classification and Framing of Educational Knowledge," in *Knowledge and Control,* 57–58.

70. D. C. Geary, *Children's Mathematical Development: Research and Practical Applications* (Washington, DC: American Psychological Association, 1994), cited in Ron Eglash, "When Math Worlds Collide: Intention and Invention in Ethnomathematics," *Science, Technology & Human Values* 22, no. 1 (1997), 89.

71. For a general discussion of this kind of pressure, see Gary Lee Downey and Juan C. Lucena, "Knowledge and Professional Identity in Engineering: Code-Switching and the Metrics of Progress," *History and Technology* 20, no. 4 (2004): 393–420. The authors describe engineering communities in different nations and identify varying emphases on personal accomplishment, national economic goals, and other indications by which technical progress is measured.

72. George Bugliarello articulated the obstacles that pragmatic, short-term goals of accrediting agencies presented for curricular reform in engineering in "Rethinking Technology: Steady State Earth or Biosoma?" (1971), UICA, 15.

73. Lawrence Grayson, *The Design of Engineering Curricula*, vol. 5, UNESCO Studies in Engineering Education (Paris: UNESCO, 1977), 60–61.
74. Paul H. Robbins, "What Engineers *Do* Is Important" (Paper presented at the American Society for Engineering Education 25th Annual College-Industry Conference, Stanford University, 1973); Engineering Manpower Commission, "Statement by the Engineering Manpower Commission of Engineers Joint Council on the Preliminary Report of the ASEE Goals Study" (1966), in IITA; see also Folk, *The Shortage of Scientists and Engineers*, 242–266.
75. National Academy of Sciences Committee on Minorities in Engineering, "Building the Multiplier Effect," 2.
76. Ibid., 7; Thomas L. Martin, Jr., speech (October 10, 1978), in IITA.
77. Robert B. Banks, "Mission of the College of Engineering," memo to Norman Parker and Glenn Terrell, August 10, 1966, in UICA; George Bugliarello, quoted in "College Holds 'University and Industry' Symposium," *Chicago Circle Engineering* 4, no. 3 (1970); Folk, *The Shortage of Scientists and Engineers*, 175–185.
78. UIC, "Recruitment, Admissions and Student Support: Background Issues" (1969), in UICA.
79. "College Holds 'University and Industry' Symposium"; Casey Banas, "At Age 10, U.I. Circle Is a Gateway of Hope," *Chicago Tribune*, March 2, 1975.
80. "An Experimental and Interdisciplinary Course: Housing for Low and Moderate Income Families," *Chicago Circle Engineering* 5, no. 3 (1971); "A Conference on—Women in Engineering: Bridging the Gap between Technology and Society," *Chicago Circle Engineering* 5, no. 3 (1971); Bugliarello and Simon, "The Impact of the University."
81. Dennis G. Daniels, "Engineering Supportive Services Program for Minority Group Students, University of Illinois at Chicago Circle" (proposal, 1972), in UICA.
82. Author interview with Richard Johnson, February 6, 2003. See also William Braden, "UICC's 'Urban Mission?': Part IV," *Chicago Chicago Sun-Times*, June 4, 1975.
83. Paul Chung, in *A History of the College of Engineering at the University of Illinois (1946–2001)*, 28, 36; UIC, "Recruitment, Admissions and Student Support."
84. "Chicago Circle Celebrates Its Tenth Anniversary," *Chicago Circle Engineering* 9, no. 1 (1975); Andy Shaw, "UICC's 'Urban Mission?': Part I," *Chicago Sun-Times*, June 1, 1975.
85. David W. Levinson, "The College of Engineering and the Community," *Chicago Circle Engineering* 2, no. 4 (1968). In congressional hearings held in 1963, employers expressed concern about the degree to which engineering graduate programs failed to instruct students in pertinent industrial applications (summarized in Folk, *The Shortage of Scientists and Engineers*, 175).
86. Levinson, "The College of Engineering and the Community." Similarly, Banks, August 10, 1966.
87. Author interviews with Richard Johnson and Fred Beuttler, March 6, 2003; Daniels, "Engineering Supportive Services Program," Appendix A1; Shaw, "UICC's 'Urban Mission?': Part I' "; Rosen, *Decision Making Chicago-Style*, 166; Author interview with Cecil Curtwright, February 6, 2003.

88. Author interview with Cecil Curtwright, February 6, 2003.
89. Engineering Manpower Commission, "Statement by the Engineering Manpower Commission of Engineers Joint Council on the Preliminary Report of the ASEE Goals Study (1966)," in IITA, 1–2.
90. Daniels, "Engineering Supportive Services."
91. Harrington, "Hilbersheimer and the Redevelopment of the South Side of Chicago," 76.
92. Slaton, *Reinforced Concrete and the Modernization of American Building, 1900–1930*.
93. Chicago Committee on Urban Opportunity, "Urban Opportunity in Chicago: A Report, 400 Days with 400,000 People" (report, March 1966), in CHM.
94. Biles, *Richard J. Daley*, 104–108.
95. Paul R. Paslay, "A Message from the Dean of the College of Engineering," *Chicago Circle Engineering* 10 (1976).
96. Folk, *The Shortage of Scientists and Engineers*, 136–137.
97. In the early 1970s, Gurin and Epps found a broad range of factors influencing those inclinations. Black students believed that jobs that did not involve being in an institutional setting, or contact with a company's clientele (as would a sales position), were less discriminatory. Further, students perceived that, "jobs that lay directly in the line of authority in production prompted more discrimination than special staff jobs outside the production line." Many technical jobs in industry match the first description, as engineering decisions generally fit tightly within design and manufacturing or construction processes. All of these perceptions were determinants for minority students choosing careers. The perceptions that discourage entry into a profession have a powerful effect, and in their smallest details can direct us to an important new understanding of exclusion. Gurin and Epps, *Black Consciousness, Identity, and Achievement*, 46–51, 64–67.
98. George Bugliarello, quoted in DeFotis, *A History of the College of Engineering at the University of Illinois (1946–2001)*, 41.
99. For an overview of sociological literature on race and occupational choice among young black and white Americans, see Hallinan, "Sociological Perspectives on Black-White Inequalities in American Schooling," and Gurin and Epps, *Black Consciousness, Identity, and Achievement*.
100. Phyllis Gillett, "Structural Unemployment: Recycling Engineers" (Paper presented at the American Society for Engineering Education 25th Annual College-Industry Conference, Stanford University, 1973), 56; Frederick E. Terman, "Manpower Projections for Engineering: The Next Quarter Century" (Paper presented at the American Society for Engineering Education 25th Annual College-Industry Conference, Stanford University, 1973), 54; John D. Alden, "Engineering Job Prospects for 1970," *Scientific Engineering Technical Manpower Comments* 7, no. 4 (1970); Juan C. Lucena, *Defending the Nation: U.S. Policymaking to Create Scientists and Engineers from Sputnik to the 'War against Terrorism'* (Lanham, MD: University Press of America, 2005), 66–67.
101. Christopher Newfield, *Ivy and Industry: Business and the Making of the American University, 1880–1980* (Durham: Duke University Press, 2003), 102.

102. Glenn C. Loury, "Foreword," in Bowen and Bok, *The Shape of the River*, xxvi.

5. Urban Engineering and the Conservative Impulse

1. As Christopher Newfield has written, "Having cloaked stratification in the languages of nature and science, meritocracy insured that future attempts to value individual difference and diversity in higher education would appear not to expand merit but to compromise it" (Newfield, *Ivy and Industry: Business and the Making of the American University, 1880–1980* [Durham: Duke University Press, 2003], 103). See also, Gary Lee Downey and Juan C. Lucena, "Knowledge and Professional Identity in Engineering: Code-Switching and the Metrics of Progress" *History and Technology* 20 (no. 4): 393–420.
2. Stuart W. Leslie, *The Cold War and American Science: The Military-Industrial-Academic Complex at MIT and Stanford* (New York: Columbia University Press, 1993), 233–256; Rebecca Slayton, "Boycotting Star Wars: A Discursive Transgression of the Science-Politics Boundary" (unpublished manuscript, 2005); Matthew Wisnioski, "Engineers and the Intellectual Crisis of Technology, 1957–1973" (Ph.D. diss., Princeton University, 2005). This was a diffuse and varied movement. Critics sometimes, but not always, distinguished between those whom they believed (innocently) produced new knowledge and those who applied it to undesirable ends: "Ballistics engineers rather than the academic mathematicians are berated for modern weapons systems (Leslie Sklair, "The Sociology of the Opposition to Science and Technology: With Special Reference to the Work of Jacques Ellul," *Comparative Studies in Society and History* 13, no. 2 [1971]: 219).
3. Because potential public sector employers had few positions for graduates of such programs, students hesitated to enroll, so few new practitioners in the field were produced to carry out its aims and create new sites of practice. Martin Jenkins, "The Urban Involvement of Higher Education" (Paper presented at the ACE Regional Conference on the Urban Involvement of Higher Education in the 1970's [Draft], Chicago, 1974), 21; Percy A. Pierre, "A Personal History" (typescript, 1998).
4. Dennis G. Daniels, "Engineering Supportive Services Program for Minority Group Students, University of Illinois at Chicago Circle" (proposal, 1972), in UICA. For an extended discussion of the political orientation of minority students, see Patricia Gurin and Edgar Epps, *Black Consciousness, Identity, and Achievement: A Study of Students in Historically Black Colleges* (New York: Wiley, 1975).
5. Sociologist Basil Bernstein articulated the close relationship of curricular, pedagogical, and evaluative practices in education: all reflect an investment in a particular set of educational identities and authority structures. This chapter approaches researchers" subjects as part of this investment. Basil Bernstein, "On the Classification and Framing of Educational Knowledge," in *Knowledge and Control,* ed. Michael F. D. Young (London: Collier-Macmillan, 1971), 57.

6. Jennifer S. Light, *From Warfare to Welfare: Defense Intellectuals and Urban Problems in Cold War America* (Baltimore: Johns Hopkins University Press, 2003).
7. George Bugliarello and Harold A. Simon, "The Impact of the University on Its Environment: The Regional Role of Engineering Colleges" (College of Engineering, University of Illinois Chicago, 1973), in UICA.
8. Thomas Gieryn, "Boundaries of Science," in *Handbook of Science and Technology Studies*, ed. Sheila Jasanoff et al. (Thousand Oaks, CA: Sage Publications, 1995), 394.
9. Wisnioski, "Engineers and the Intellectual Crisis of Technology, 1957–1973."
10. On the compelling nature of associations of "logical" thought with science and "nonlogical" thought with social sentiments, see M. D. King, "Reason, Tradition and the Progressiveness of Science," *History and Theory* 10, no. 1 (1971): 4–5.
11. David Noble, *America by Design: Science, Technology and the Rise of Corporate Capitalism* (Oxford: Oxford University Press, 1977), 20–32, 317–318; Newfield, *Ivy and Industry*; Amy E. Slaton, *Reinforced Concrete and the Modernization of American Building, 1900–1930* (Baltimore: Johns Hopkins University Press, 2001).
12. For contextualization of this trend, see Kathryn A. Neeley, "Liberal Studies and the Integrated Engineering Education of ABET 2000: Reports from a Planning Conference at the University of Virginia, April 4–6, 2002."
13. Lawrence P. Grayson, *The Design of Engineering Curricula*, vol. 5, UNESCO Studies in Engineering Education (Paris: UNESCO, 1977), 55.
14. IIT Institutional Study Committee, "Report of the Institutional Study Committee of Illinois Institute of Technology on Goals of Engineering Education/American Society for Engineering Education" (report, 1964), in IITA, 4; see also Slaton, *Reinforced Concrete and the Modernization of American Building, 1900–1930*.
15. David G. Hammond, "Matching Engineering Solutions to Society's Problems: BART—A Case History" (Paper presented at the American Society for Engineering Education 25th Annual College-Industry Conference, Stanford University, 1973), 106. This is a remarkably popular locution.
16. Omi and Winant describe the "long hot summers" of the mid-1960s as striking many as a "proto-revolutionary situation." Michael Omi and Howard Winant, *Racial Formation in the United States*, 2d ed. (New York: Routledge, 1994), 196 n.22; also, Jenkins, "The Urban Involvement of Higher Education."
17. Jenkins, "The Urban Involvement of Higher Education." At least one UIC vice chancellor, Michael Goldstein, was deeply disappointed in the ACE conference, finding the university leaders who attended to be uninspired. But the conferences did attract many prominent educators, mayors, and governors. Michael Goldstein to Martin Jenkins, April 22, 1974, in UICA.
18. Slayton, "Boycotting Star Wars"; Wisnioski, "Engineers and the Intellectual Crisis of Technology, 1957–1973."
19. Hammond, "Matching Engineering Solutions to Society's Problems," 108.

20. Daniels, "Engineering Supportive Services Program for Minority Group Students," 6; Paul R. Paslay, "A Message from the Dean of the College of Engineering," *Chicago Circle Engineering* 10 (1976). IIT's 1974 Board of Trustees report claimed, simply, that "[t]he need for centers of research and scholarship focused on the relationship of science and technology to social and human needs has never been greater" (IIT, Board of Trustees, "The Future for Illinois Institute of Technology" [report, 1974], in IITA, 2).
21. Jim Jenson, "Press Release" (1951), in IITA. Chicago's prominence in metals; non-electrical machinery; and stone, clay, and glass products provided consulting opportunities of which IIT faculty took great advantage. In 1970, 54 percent of the consulting done by IIT faculty was done for Chicago-based clients. Bugliarello and Simon, "The Impact of the University," 54–59.
22. Christopher Newfield describes this trend: "Customized, pricey corporate seminars are likely to tell their paying customers what they want to hear"; Newfield, "Jurassic U: The State of University-Industrial Relations," *Social Text* 22, no. 2 (2004): 40. See also Sheila Slaughter and Larry L. Leslie, *Academic Capitalism: Politics, Policies, and the Entrepreneurial University* (Baltimore: Johns Hopkins University Press, 1997).
23. Chicago Department of Development and Planning, "The Comprehensive Plan of Chicago, Conditions and Trends: Population, Economy, Land" (Staff Study Report, 1967), in CHM, 77.
24. The University of Pittsburgh had founded the Mellon Institute in 1913, beginning a trend that slowly but consistently grew among U.S. universities. See Newfield, *Ivy and Industry*, 37.
25. Bugliarello and Simon, "The Impact of the University," 68–72; Author interview with Sy Bortz, May 17, 2004; IIT, Board of Trustees, "The Future for Illinois Institute of Technology" [report, 1974], in IITA, 2–3; Leslie, *The Cold War and American Science*, 243.
26. Bugliarello and Simon, "The Impact of the University."
27. Ibid., 27, 33–35, 97.
28. Leslie, *The Cold War and American Science*, 2.
29. Similarly conspicuous protests around the country over the next four or five years melded concerns about the Vietnam War, capitalism, and environmental degradation in various combinations. See Wisnioski, "Inside 'The System,'" 24–31; Leslie, *The Cold War and American Science*, 239.
30. IIT, "IIT Academic Catalogue," 1969, in IITA.
31. Bugliarello and Simon, "The Impact of the University," 82–83.
32. Author interview with Richard Johnson, February 6, 2003; William Braden, "UICC's 'Urban Mission?': Part IV," *Chicago Sun-Times*, June 4, 1975; Paul Chung, in William DeFotis, *A History of the College of Engineering at the University of Illinois (1946–2001)*, (Chicago: University of Illinois at Chicago, 2001), 36.
33. Bugliarello and Simon, "The Impact of the University," 49–50.
34. "Chicago Circle Celebrates Its Tenth Anniversary," *Chicago Circle Engineering* 9, no. 1 (1975); Andy Shaw, "UICC's 'Urban Mission?': Part I," *Chicago Sun-Times*, June 1, 1975.

35. DeFotis, *A History of the College of Engineering at the University of Illinois (1946–2001)*, 17, 36.
36. UIC, "UICC 1979 Mission Statement" (1980), in UICA, 15–19.
37. Illinois Board of Higher Education, "Master Plan for Illinois Higher Education: Report of Committee V, Graduate Education in Engineering" (1970), in UICA, 7–15.
38. University of Illinois, "Commentary on Master Plan-Phase III, Initial Draft" (1971), in UICA, 16. Tellingly, UIC Dean of Faculties Terrell had once indicated that a university-run preschool for inner-city children carried the most import for social change among the university's activities: "I do not believe the University could indicate its interest in the proposal of urban society [sic] any more significantly than through the support of projects like this" (Glenn Terrell to Nan E. McGehee and Robert E. Corley, Committee on Human Relations and Equal Opportunity, November 5, 1965, in UICA). Educational interventions at the college level are secondary, at best, in this outlook.
39. Braden, "UICC's 'Urban Mission?': Part IV"; Reynold Feldman, "Conference/Travel Report: The ACE Regional Conference on the Urban Involvement of Higher Education in the 1970s" (1974), in UICA, 2; Newfield, *Ivy and Industry*, 176–177.
40. Illinois Board of Higher Education, "Master Plan: Engineering (Committee V): Draft" (1970), I-2–I-3.
41. Fred W. Beuttler, "Envisioning an Urban University: President David Henry and the Chicago Circle Campus of the University of Illinois, 1955–1975," *History of Education Annual* (2003): 22–23.
42. Steven B. Sample, "The Evolution of Engineering Education," *Chicago Circle Engineering* 6, no. 3 (1972); Shaw, "UICC's 'Urban Mission?': Part I."
43. For interesting exceptions, see American Society for Engineering Education, "Liberal Learning for the Engineer: Report of the ASEE Humanistic-Social Research Project" (Washington, DC: 1968), 36–37.
44. George Bugliarello, "Rethinking Technology: Steady State Earth or Biosoma?" (1971), in UICA.
45. Bugliarello and Simon, "The Impact of the University," 16–17. Bugliarello recognized, too, that it was entrance requirements for graduate school that pushed "sensitivity to social problems" out of the undergraduate engineering curriculum in favor of a mere "compressing of additional new knowledge" (p. 16).
46. Clarence G. Williams, *Technology and the Dream: Reflections on the Black Experience at MIT, 1941–1999* (Cambridge, MA: MIT Press, 2001), 104–107.
47. Bugliarello and Simon, "The Impact of the University," 7–9.
48. Theodore M. Porter, *Trust in Numbers: The Pursuit of Objectivity in Science and Public Life* (Princeton, NJ: Princeton University Press, 1995), 149; Amy Slaton and Janet Abbate, "The Hidden Lives of Standards: Technical Prescriptions and the Transformation of Work in America," in *Technologies of Power: Essays in Honor of Thomas Parke Hughes and Agatha Chipley Hughes*, ed. Michael Thad Allen and Gabrielle Hecht (Cambridge, MA: MIT Press, 2001): 95–143.

49. Lawrence R. Holt and Tom Lewis, "Divided Highways: The Interstates and the Transformation of American Life" (film), 1997.
50. Bugliarello and Simon, "The Impact of the University," 75, 94.
51. Hammond, "Matching Engineering Solutions to Society's Problems," 109; University of Illinois, "Commentary on Master Plan-Phase III," 16, 24–27.
52. See Steve Woolgar, "Introduction: Technology and Society," in *Science, Technology and Society: An Encyclopedia*, ed. Sal Restivo (Oxford: Oxford University Press, 2005), xix–xx.
53. Gail Dubrow and Mary Corbin Sies, "Letting Our Guard Down: Race, Class, Gender and Sexuality in Planning History," *Journal of Planning History* 1, no. 3 (2002): 203–214; Kelly Quinn, "Planning History/Planning Race, Gender, Class and Sexuality," *Journal of Planning History* 1, no. 3 (2002), 241.
54. Chicago Department of Development and Planning, "The Comprehensive Plan of Chicago."
55. Philip Mirowski, "Cyborg Agonistes: Economics Meets Operations Research in Mid-Century," *Social Studies of Science* 29, no. 5 (1999): 694–695.
56. Jonathan Cohn, "Irrational Exuberance: When Did Political Science Forget About the Politics?" *The New Republic* (October 25, 1999): 25–31.
57. Karin Knorr-Cetina, "Laboratory Studies: The Cultural Approach," in *Handbook of Science and Technology Studies*, ed. S. Jasanoff et al. (Thousand Oaks, CA: Sage Publications, 1995): 140–166.
58. Editors at the *Chicago Tribune* lavishly praised Rettaliata's handling of this incident; Thomas Powers, "Rettaliata's 'No' Saved I.I.T.'S Peace," *Chicago Tribune*, July 5, 1970.
59. Daniel P. Moynihan, "To Solve Problem, First Define It," *New York Times*, January 12, 1970. This quote is lauded by John Connolly, Illinois State Representative, in an address to the UIC College of Engineering; John H. Conolly, "Needed: An Interface between Science and Politics," *Chicago Circle Engineering* 7, no. 1 (1973).
60. John T. Rettaliata, "Speech Delivered on May 22, 1952," in IITA, 1–2.
61. John D. Kemper, "The Engineering Outlook: Shortage or Surplus?" (Paper presented at the American Society for Engineering Education 25th Annual College-Industry Conference, Stanford University, 1973), 59.
62. Bugliarello and Simon, "The Impact of the University," 59.
63. Cited in ibid., 37.
64. Editorial, "Chicago's Stake in IIT," *Chicago Daily News*, May 28, 1965. Some Americans may understandably have associated industrial enterprise with an oppressive state apparatus on hearing President Johnson's conceptions of how corporations might help suppress unrest. In 1968, Johnson told corporate representatives that if the "hard-core unemployed" of the ghetto are working, "they won't be throwing bombs in your homes and plants." Terry H. Anderson, *The Pursuit of Fairness: A History of Affirmative Action* (Oxford, Oxford University Press, 2004), 105.
65. Neeley, "Liberal Studies and the Integrated Engineering Education of ABET 2000," 14–16; Slaton, *Reinforced Concrete and the Modernization of American Building*, 52–54.

66. Rettaliata, "Speech Delivered on May 22, 1952," in IITA, 3–4.
67. IIT, Board of Trustees, "The Future for Illinois Institute of Technology," 3.
68. An entire graduate program at Princeton, established in the mid-1960s, offered doctoral work in "Humanistic Studies in Engineering" and promised to fit engineering courses "directly into a cultural perspective." That perspective was constituted of close structural analysis of architecture to reveal the aesthetic and cultural achievements of engineers, arguably a narrow definition of culture. Similar courses are being taught at Princeton today. American Society for Engineering Education, "Liberal Learning for the Engineer," 37.
69. Bugliarello, "Rethinking Technology."
70. Ibid.; Porter, *Trust in Numbers*.
71. American Society for Engineering Education, "Liberal Learning for the Engineer."
72. George Bugliarello and Dean B. Doner, "Editors' Preface," in *The History and Philosophy of Technology* (Urbana, IL: University of Illinois Press, 1979).
73. Ibid., x.
74. Juan C. Lucena, *Defending the Nation: U.S. Policymaking to Create Scientists and Engineers from Sputnik to the "War against Terrorism"* (Lanham, MD: University Press of America, 2005).
75. John D. Alden, "Engineering Job Prospects for 1970," *Scientific Engineering Technical Manpower Comments* 7, no. 4 (1970).
76. Newfield, *Ivy and Industry*, 217–218.
77. Bruce Seely, "The Other Re-Engineering of Engineering Education, 1900–1965," *Journal of Engineering Education* (1999): 288–290.

6. Race and the New Meritocracy

1. "Student Attention Goal of President of Prairie View," *Battalion*, November 2, 1977.
2. Amilcar Shabazz, *Advancing Democracy: African Americans and the Struggle for Access and Equity in Higher Education in Texas* (Chapel Hill: University of North Carolina Press, 2004); Alexander Astin, "The Myth of Equal Access in Public Higher Education," in *Southern Educational Foundation* (1975), cited in Lorenzo Morris, *Elusive Equality: The Status of Black Americans in Higher Education* (Washington, DC: Howard University Press, 1979), 56–57.
3. Through the twentieth century, PVAMU produced the vast majority of Texas' black teachers and nurses, and in some decades more engineers than any other historically black college or university, but it did so with drastically underfunded facilities. For an overview of the school's status at mid-century, see West A. Hamilton, "Prairie View Has Become Negro Educational and Cultural Capital of Texas," *Houston Chronicle*, August 3, 1952. See also Willie Pearson Jr. and LaRue C. Pearson, "Baccalaureate Origins of Black American Scientists: A Cohort Analysis," *JNE* 54, no. 1 (1985), cited in Elaine P. Adams, "Benjamin Banneker Honors College: Gateway to Scientific and Technical Doctorates," *JNE* 59, no. 3 (1990): 452; Joshua Rolnick, "Minority Enrollment Drops in Graduate Science Programs," *CHE* 45, no. 4 (1998): A48. Black participation in higher education was dropping overall in this period, as backlash against

civil rights developments undermined federal and state commitments to the HBCUs; see Adams, "Benjamin Banneker Honors College," 449–450; Manning Marable, "The Beast Is Back: An Analysis of Campus Racism," *The Black Collegian*, September/October (1990): 53, cited in Evonne Parker Jones, "The Impact of Economic, Political, and Social Factors on Recent Overt Black/White Racial Conflict in Higher Education in the United States," *JNE* 60, no. 4 (1991): 524–525. Lorenzo Morris describes the rise of anti-affirmative action sentiments after the *Bakke* decision in Morris, *Elusive Equality*, 3. See also Philip G. Attach and Koki Looter, eds., *The Racial Crisis in American Higher Education* (Albany: State University of New York Press, 1991); Gary Orfield and Faith Paul, "Declines in Minority Access: A Tale of Five Cities," *Educational Record* 68, no. 4 and *Educational Record* 69, no. 1(1987–1988): 59–62.

4. Frances R. Aparicio, "On Multiculturalism and Privilege: A Latina Perspective," *American Quarterly* 46, no. 4 (1994): 575–588; W.G Bowen and D. Bok, *The Shape of the River: Long-Term Consequences of Considering Race in College and University Admissions* (Princeton, NJ: Princeton University Press, 1998); Morris, *Elusive Equality*, 3. On the "miniscule" representation of minority practitioners in graduate level science, engineering, and mathematics, see Adams, "Benjamin Banneker Honors College"; and Ronald Roach, "Losing Ground," *BIHE* 21, no. 2 (2004): 28–29. In 1995, African Americans, Hispanics, and Native Americans accounted for 28 percent of the college-age population, yet earned only 9 percent of bachelor's degrees that year; Beth Panitz, "Policy Shifts," *ASEE Prism* 6 (1996): 15–16.
5. Rolnick, "Minority Enrollment Drops in Graduate Science Programs."
6. Robin J. Ely and David A. Thomas, "Cultural Diversity at Work: The Effects of Diversity Perspectives on Work Group Processes and Outcomes," *Administrative Science Quarterly* 46, no. 2 (2001): 229–273; Avery Gordon, "The Work of Corporate Culture: Diversity Management," *Social Text* 13, no. 3 (1995): 3–30; Jane Juffer, "The Limits of Culture," *Neplanta* 2, no. 2 (2001): 265–273; Christopher Newfield, *Ivy and Industry: Business and the Making of the American University, 1880–1980* (Durham: Duke University Press, 2003).
7. State of Texas, "Texas Equal Educational Opportunity Plan for Higher Education" (1983), 2.
8. The "Key Thesis" is described in Chandler Davidson, *Race and Class in Texas Politics* (Princeton, NJ: Princeton University Press, 1990), 3–7; Shabazz, *Advancing Democracy*; Jama Lazerow, "The Multiple Meanings of Advancing Democracy [Book Review]," *Reviews in American History* 32, no. 3 (2004): 422–430. Lazerow, summarizing Shabazz, says: "Texas was the site of a decisive shift in civil rights activism: from legal demands for separate-but-equal facilities to a direct challenge to the whole system of racially separate education" (p. 426).
9. On the general tendency of race-blind programs to fail in this regard, see Shirley M. Malcom, Daryl E. Chubin, and Jolene K. Jesse, *Standing Our Ground: A Guidebook for STEM Educators in the Houston Post-Michigan Era* (Washington, DC: American Association for the Advancement of Science, 2004), 33–34; "Hopwood Tests Mettle," *Austin American Statesman*, February 20, 1997;

Marta Tienda et al., *Closing the Gap?: Admissions and Enrollments at the Texas Public Flagships before and after Affirmative Action* (Woodrow Wilson School of Public and International Affairs, Princeton University, 2003), 31. A federal task force in 1989 found that while blacks comprised only 2 percent of all employed scientists and engineers in the United States (despite comprising 12 percent of the general population), most blacks holding advanced degrees completed their degrees at HBCUs; Adams, "Benjamin Banneker Honors College," 450. This situation meant that PVAMU's potential to contribute to the ranks of degree-holding black engineers was considerable; see Raymond Richardson, quoted in Janita Poe, "Traditional Black Colleges Struggle to Create Diversity While Preserving Proud Histories," *Atlanta Journal-Constitution*, November 4, 2001, 1A.

10. See Samuel DuBois Cook, "The Socio-Ethical Role and Responsibility of the Black-College Graduate," in *Black Colleges in America: Challenge, Development, Survival*, ed. Charles V. Willie and Ronald R. Edmonds (New York: Teachers College Press, 1978); S. Steele, *The Content of Our Character: A New Vision of Race in America* (New York: St. Martin's Press, 1990). For an exemplary analysis of the uses of diversity in U.S. corporations, see Gordon, "The Work of Corporate Culture."

11. Chester M. Hedgepeth Jr., Ronald R. Edmonds, and Ann Craig, "Overview," in *Black Colleges in America*, 18. Charles Willie sees Benjamin Mayes, president of Morehouse College, as typical of HBCU presidents in defining education there "as a moral experience that motivates one to be concerned about others, especially those who are oppressed and treated unjustly." Charles V. Willie, "Racism, Black Education, and the Sociology of Knowledge," in *Black Colleges in America*, 13. See also Walter R. Allen and Joseph O. Jewell, "A Backward Glance Forward: Past, Present, and Future Perspectives on Historically Black Colleges and Universities," *The Review of Higher Education* 25, no. 3 (2002): 241–261.

12. Texas Higher Education Coordinating Board, *Priority Plan to Strengthen Education at Prairie View A&M University and at Texas Southern University* (Texas Higher Education Coordinating Board, Austin, TX.: 2000), i, 84–85, 92; John B. Williams, "Systemwide Title VI Regulation of Higher Education, 1968–88: Implications for Increased Minority Participation," in *The Education of African-Americans*, ed. Charles V. Willie, Antoine M. Garibaldi, and Wornie L. Reed (New York: Auburn House, 1991: 110–118; Bowen and Bok, *The Shape of the River*, 10–13.

13. Tom Nelson and David Ellison, "Prairie View Begins Tough Task of Improving," *Houston Post*, May 15, 1983.

14. There were also six private black colleges in the state; see Vernon McDaniel, "Negro Publicly-Supported Higher Institutions in Texas," *JNE* 31, no. 3 (1962): 349–353.

15. For the sake of simplicity, I will refer to the institution of higher learning at Prairie View as PVAMU, despite its frequent change of official title through the middle of the twentieth century.

16. Orfield and Paul, "Declines in Minority Access."

17. PVAMU, Office of Institutional Research, "Facts About... Prairie View A&M College of Texas" (Prairie View, TX: Prairie View A&M College, 1968), 8; Edmund W. Gordon, *A Descriptive Analysis of Programs and Trends in Engineering Education for Ethnic Minority Students (Institution for Social and Policy Studies of Yale University)* (New York: National Action Council for Minorities in Engineering, 1986), 4.

18. Regarding the role of civil engineers in Texas' economic development, we might consider that by 1950 the state held more miles of interstate highway than any other state (3,215 miles), and over 70,000 miles of paved roads. See Joseph E. King, "Civil and Mechanical Engineering," in *100 Years of Science and Technology in Texas: A Sigma Xi Centennial Volume*, ed. Leo Klosterman (Houston: Rice University Press, 1986), 207.

19. Federal Reserve Bank of Dallas, "Don't Mess with Texas," *Southwest Economy*, Issue 1 (January/February 2005), at www.dallasfed.org (accessed 6/22/2009); see also Lewis Harris, Phillip Hopkins, and Sue A. Krenek, "Computers," in *100 Years of Science and Technology in Texas*; Donald Lyons and Bill Luker Jr., "Employment in R & D-Intensive High-Tech Industries," *Monthly Labor Review* (November 1996): 15–25.

20. Other engineering colleges in the state followed this same path. In 1924, the University of Texas set up the Bureau of Engineering Research, which became an important center of scientific research and development during World War II; "UT Carries on National Research," *Texas Professional Engineer* 5, no. March–April (1946): 9; Texas Tech, a smaller institution, established the Institute for Disaster Research in the early 1970s, and all of the schools helped sustain local and national professional engineering organizations; King, "Civil and Mechanical Engineering."

21. PVAMU, College of Engineering, *Report of the College of Engineering for the Prairie View A&M University Long-Range Development Plan 1981–1987* (Prairie View, TX: Prairie View A&M University, 1980), 7–13; Shabazz, *Advancing Democracy*, 19; Jane E. Smith Browning and John B. Williams, "History and Goals of Black Institutions of Higher Learning," in *Black Colleges in America*, 91.

22. Shabazz, *Advancing Democracy*, 23.

23. Browning and Williams, "History and Goals of Black Institutions of Higher Learning," 91; Hamilton, "Prairie View Has Become Negro Educational and Cultural Capital of Texas"; McDaniel, "Negro Publicly-Supported Higher Institutions in Texas," 351; PVAMU, School of Engineering, "Dean C. L. Wilson Biography (Dedication and Awards Ceremony)," in PVAMUA.

24. "Students Learn Building Methods at Prairie View," *Waco Tribune*, December 12, 1951; "Wilson [Obituary]," *Houston Post*, January 26, 1994; McDaniel, "Negro Publicly-Supported Higher Institutions in Texas"; PVAMU, Office of Institutional Research, "Facts About... Prairie View A&M College of Texas"; Nelson and Ellison, "Prairie View Begins Tough Task of Improving." On the role of technology research in institutional growth, see Bruce Seely, "Research, Education and Science in American Engineering Colleges: 1900–1960," *Technology and Culture* 34, no. 2 (1993): 344–386; Amy E. Slaton, "George Washington

Carver Slept Here: Racial Identity and Laboratory Practice at Iowa State College," *History and Technology* 17 (2001): 353–375.
25. PVAMU, College of Engineering, "Report of the College of Engineering for the Prairie View A&M University Long-Range Development Plan 1981–1987," 17, 21.
26. Governor John Connelly, "Charge to the Coordinating Board Texas College and University System," (1965); John R. Thelin, "Looking for the Lone Star Legacy: Higher Education in Texas," *History of Education Quarterly* 17, no. 2 (1977), 222.
27. Governor Connelly, "Charge to the Coordinating Board Texas College and University System"; Author interview with Kenneth Ashworth, July 7, 2005.
28. PVAMU, College of Engineering, "Report of the College of Engineering for the Prairie View A&M University Long-Range Development Plan 1981–1987," 18. This kind of outreach was customary for many of the PVAMU faculty. Associate Professor Sam R. Daruvalla, when appointed chairman of the educational and student affairs committee of the Houston Section of the Institute of Electrical and Electronics Engineers (IEEE), amalgamated student chapters at PVAMU, TAMU, University of Houston, and Rice University; "Prairie View Professor Named Coordinator," *Bryan-College Station Eagle*, September 17, 1975.
29. "Richard Price Biography," at jrank.biography.org (accessed 11/1/08). The University of Texas was viewed by other schools as likely to draw off "the cream of the academic crop." Edward Sheridan, quoted in Lydia Lum, "Is It Just a Campaign 'Commitment'?" *BIHE* 22, no. 9 (2000): 24.
30. "Dr Thomas New Prairie View President," *Houston Post,* November 24, 1966; PVAMU, School of Engineering, "Southern Association of Colleges and Secondary Schools Departmental Self-Study," August, 1968, in PVAMU, 8; Austin E. Greaux, "The Engineering Profession and the Negro" (Paper presented to the Sam Houston Chapter of the Texas Society of Professional Engineers, March 21, 1968).
31. PVAMU, College of Engineering, "Report of the College of Engineering for the Prairie View A&M University Long-Range Development Plan 1981–1987," I, 30, 71–73, 161; "Energy Affair Center Approved," *Bryan-College Station Eagle*, December 9, 1980. Some initiatives began in non-engineering technical programs at PVAMU; through the early 1970s, the School of Industrial Education and Technology, in close collaboration with large corporations such as the Dow Chemical Company of Freeport, Texas, set up off-campus programs that carried over 200 Prairie View students to companies around the sate. Other students worked parttime through the school year at U.S. Steel, General Electric, and Ford plants around the country. See PVAMU, School of Industrial Education and Technology, "Highlights of Accomplishments in 1972–73", in "Annual Report" (1972–1973), in PVAMUA. PVAMU's emerging engineering departments rapidly built on such connections, dispatching dozens of students to co-op positions at many of the same firms, and at such government agencies as the National Bureau of Standards and NASA. "Prairie View A&M Active in 'Co-Op'," *Bryan-College Station Eagle,* October 15, 1980.

32. PVAMU, College of Engineering, "Report of the College of Engineering for the Prairie View A&M University Long-Range Development Plan 1981–1987," 191–192.
33. Saralee Tiede, "Black Universities Pose Tough Problem for State Planners," *Bryan-College Station Eagle*, September 28, 1985.
34. On the fluctuating fortunes of Texas A&M System schools in regard to state funding (and competition with College Station), including distributions of funding to Tarleton State University, heavily attended by Hispanic students, see "Williams Rebuts Charge of Prairie View Neglect," *Battalion*, September 7, 1978; Jim Davis, "Caucus Seeks Ruling on Funds for Prairie View," *Bryan-College Station Eagle*, January 11, 1980; W. R. Deener III, "Prairie View to Push for Share from Fund," *Dallas News*, December 19, 1982; Fred Bonavita, "Prairie View Controversy: Delco Accuses Attorney General of Avoiding Sensitive Issue," *Houston Post*, February 28, 1982; Tom Nelson, "Officials Want Doubling of Prairie View Funds," *Houston Post*, September 2, 1982; and Tom Nelson, "Prairie View A&M's Frayed Facilities in Need of a Long Neglected Facelift," *Houston Post*, May 15, 1983.
35. Nelson, "Prairie View A&M's Frayed Facilities in Need of a Long Neglected Facelift"; Tom Nelson, "Prairie View A&M May Get President by First of Year," *Houston Post*, November 24, 1982; Percy A. Pierre and Elaine P. Adams, "Combating Minority Brain Drain," *Educational Record* 68, no. 2 (1987): 48–53.
36. In 1982, 70 percent of students enrolled at Prairie View came from families with annual incomes below $8,000; of 4,661 students, 3,400 received financial aid. See Nelson, "Officials Want Doubling of Prairie View Funds."
37. PVAMU, "Prairie View A&M University Projected Facilities, 1982–2000" (1980), in *A. I. Thomas Papers*, PVAMUA; "UT, A&M Offer Funds to Prairie View," *Dallas Times Herald*, December 22, 1982.
38. Hansen quoted in Nelson, "Prairie View A&M's Frayed Facilities in Need of a Long Neglected Facelift"; Steve Vinson, "Poor Cash Management Allowed Prairie View to Misspend PUF," *Bryan-College Station Eagle*, August 19, 1987. Some believed that A&M officials tolerated "much lower management standards from Prairie View officials than they demanded from the rest of the system"; Nelson, "Prairie View A&M's Frayed Facilities in Need of a Long Neglected Facelift." Proponents of HBCUs pointed out that if Jencks and Reisman were right when they published their strong criticisms of HBCUs in 1967, how could these schools have improved so notably in the following decade? See Charles V. Willie and Marlene Y. MacLeish, "The Priorities of Presidents of Black Colleges," in *Black Colleges in America*.
39. Nelson and Ellison, "Prairie View Begins Tough Task of Improving"; Christopher Jencks and David Riesman, "The American Negro College," *Harvard Educational Review* 37 (1967): 48–49, cited in Willie and MacLeish, "The Priorities of Presidents of Black Colleges," 133.
40. Scot K. Meyer, "Prairie View Funds Deficient," *Battalion*, August 13, 1980.
41. Critical assessments of PVAMU arose from a range of sources, and it is often difficult to ascertain their purpose. For some, any criticisms of HBCUs gave

unwanted fuel to their opponents. Others believed that without harsh, incisive critique to expose the schools' shortcomings, no strong proof could be offered of their historic neglect. See Jacqueline Fleming, *Blacks in College: A Comparative Study of Students' Success in Black and White Institutions* (San Francisco: Jossey-Bass, 1984).

42. Thomas often channeled state funds away from building construction because he felt he could equip and furnish the buildings through private contributions. See Nelson and Ellison, "Prairie View Begins Tough Task of Improving," 1.

43. Nelson, "Prairie View A&M May Get President by First of Year"; Nelson, "Prairie View A&M's Frayed Facilities in Need of a Long Neglected Facelift"; Nelson and Ellison, "Prairie View Begins Tough Task of Improving." Similarly, author interview with Kenneth Ashworth, July 7, 2005.

44. Lorenzo Middleton, "Government to Help Black Colleges that Help Themselves, Official Says," *CHE* (1979): 9. Carter's "Memorandum" of January 25, 1978, is reproduced in J. Christopher Lehner, *Losing Battle: The Decline in Black Participation in Graduate and Professional Education* (Washington, DC: National Advisory Committee on Black Higher Education and Black Colleges and Universities, 1981).

45. Texas Higher Education Coordinating Board, ed., *Texas Equal Educational Opportunity Plan for Higher Education* (1981 [1983]), 94; Felton West, "Clements Won't Seek Funds for TSU in Special Session," *Houston Post*, January 4, 1980, 4A.

46. Fred King, "Officials Hope for OK on Racial Compliance in Texas Universities," *Houston Post*, January 2, 1981.

47. "UT, A&M Offer Funds to Prairie View"; Nelson, "Officials Want Doubling of Prairie View Funds." Hansen later summarized the situation at PVAMU as "the most confused and difficult situation I had encountered," quoted in Nelson, "Prairie View A&M's Frayed Facilities in Need of a Long Neglected Facelift."

48. Texas Higher Education Coordinating Board, "The Texas Plan for Equal Educational Opportunity: A Brief History," at www.thecb.state.tx.us (accessed 7/3/2005); Pierre and Adams, "Combating Minority Brain Drain"; Texas Higher Education Coordinating Board, "Report of the Coordinating Board Pursuant to House Resolution 527 on Progress Achieved in the Enhancement of Facilities and Programs at Texas Southern University and Prairie View A&M University through December 1, 1983, under the Texas Equal Educational Opportunity Plan for Higher Education" (1983), in PVAMUA. Author interview with Percy Pierre, July 26, 2005.

49. Nelson became executive assistant to Hansen at the system headquarters in College Station, and later president of Central Washington and Lincoln Universities. See "President Ivory V. Nelson's Profile" at www.lincoln.edu (accessed 11/30/08); David Ellison, "New Prairie View President Tabs Funding as Top Priority," *Houston Post*, January 29, 1983.

50. Author interview with Percy Pierre, July 26, 2005; Adams, "Benjamin Banneker Honors College," 452.

51. Author interview with Percy Pierre, July 26, 2005. Established patterns of graduate school enrollment for PVAMU engineering students (around 15 percent as

of 1968) provided a basis for this expansion. See PVAMU, School of Engineering, "Southern Association of Colleges and Secondary Schools Departmental Self-Study," August, 1968, in PVAMUA.

52. Pierre and Adams, "Combating Minority Brain Drain," 51.
53. Ibid. Also, Adams, "Benjamin Banneker Honors College," 450; Fritz Lanham, "Pierre Wants Whiz Kids," *Bryan-College Station Eagle,* November 23, 1983. Pierre sees curricular innovations at Florida A&M, another HBCU, under Fred Humphries as perhaps the closest to his own in this period. Author interview with Percy Pierre, July 26, 2005.
54. In 1989, 45 percent of BBHC students were from Texas, 33 percent from other states, and the rest from abroad. See Adams, "Benjamin Banneker Honors College," 456–457.
55. This might be considered one resolution of old tensions among disciplines; Willie and MacLeish, "The Priorities of Presidents of Black Colleges," 143.
56. Author interview with Jewel Prestage, March 24, 2006. The college's advisory board held members from around the nation; "Banneker Advisory Board Meets," *Banneker Intercom* 1, no. 5 (n.d.): 1–2. Michael N. Bastedo and Patricia J. Gumport, "Access to What? Mission Differentiation and Academic Stratification in U.S. Public Higher Education," *Higher Education* 46, no. 3 (2003): 352.
57. PVAMU, "Benjamin Banneker Honors College, Prairie View A&M University" [brochure, c. 1989], in PVAMUA; Ronald J. Sheehy, "The Benjamin Banneker Honors College: A Model for the Future," *National Honors Report* 8, no. 1 (1987); Pierre and Adams, "Combating Minority Brain Drain"; Willie and MacLeish, "The Priorities of Presidents of Black Colleges," 142; Terese Loeb Kreuzer, "The Bidding War for Top Black Students," *JBHE,* no. 2 (1993–1994: 114–118. On the legacy of black colleges as means of enrolling students who might otherwise not attend colleges, see Gregory Kannerstein, "Black Colleges: Self-Concept," in *Black Colleges in America,* ed. Charles V. Willie and Ronald R. Edmonds.
58. Morris, *Elusive Equality,* 200. Democratic Representative Foster Whaley proposed that the maintenance of TSU and the University of Houston upheld Jim Crow policies, and was even reminiscent of Nazi Germany. See "Closing of TSU, Prairie View to Be Urged in Bill," *Houston Post,* January 7, 1981. Kenneth Ashworth recalled supporters of the two black schools singing "We Shall Overcome" at a hearing regarding a merger of the institutions (author interview, July 17, 2005). See also Evonne Parker Jones's discussion of Steele, *The Content of Our Character,* in "The Impact of Economic, Political, and Social Factors," 533.
59. PVAMU, "A Plan to Increase the Number of White Students Attending Prairie View A&M University" (1981), in A. I. *Thomas Papers,* PVAMUA; Brad Owens, "Prairie View Reports Black Enrollment Drop," *Bryan-College Station Eagle,* January 30, 1986.
60. Owens, "Prairie View Reports Black Enrollment Drop"; PVAMU, "A Plan to Increase the Number of White Students Attending Prairie View A&M University." Houston has grown steadily toward the northwest since the late 1980s, in the direction of Prairie View. This has likely encouraged commuting

patterns, according to TAMU dean Dick Perry (author interview, June 5, 2002).

61. Author interview with Percy Pierre, July 26, 2005. See Clarence G. Williams, *Technology and the Dream: Reflections on the Black Experience at MIT, 1941–1999* (Cambridge, MA: MIT Press, 2001); Joe R. Feagin, Hernan Vera, and Nikitah Imani, *The Agony of Education: Black Students at White Colleges and Universities* (New York: Routledge, 1996).
62. "Student Attention Goal of President of Prairie View."
63. Ibid.
64. Ibid. See also K. G. Mommsen, "Professionalism and the Racial Context of Career Patterns among Black American Doctorates: A Note on the 'Brain Drain' Hypothesis" (1972), cited in Patricia Gurin and Edgar Epps, *Black Consciousness, Identity, and Achievement: A Study of Students in Historically Black Colleges* (New York: Wiley, 1975), 65.
65. Gurin and Epps, *Black Consciousness, Identity, and Achievement*, 8.
66. Ibid., 2. See also Ronald J. Sheehy, "A Significant Creation in Education," *Houston Chronicle*, October 27, 1986; author interview with Percy Pierre, July 26, 2005.
67. Adams, "Benjamin Banneker Honors College," 455.
68. Benjamin Banneker Honors College, *Annual Report* (Prairie View, TX: Prairie View A&M University, 1994–1995), 3.
69. Author interviews with Jewel Prestage, March 24, 2006; Percy Pierre, July 26, 2005; and Amilcar Shabazz, January 18, 2007.
70. Benjamin Banneker Honors College, *Annual Report* (Prairie View, TX: Prairie View A&M University, 1994–1995), 14. Things had not improved a year later: Benjamin Banneker Honors College, *Annual Report* (Prairie View, TX: Prairie View A&M University, 1995–1996), 19.
71. Author interviews with Percy Pierre, July 26, 2005, and Jewel Prestage, March 24, 2006; "Texas Commitment to Prairie View A&M University FY 2002–2003 OCR Priority Plan" at www.pvamu.edu (accessed 9/1/06), 67.
72. Author interview with Amilcar Shabazz, January 18, 2007; Willie and MacLeish, "The Priorities of Presidents of Black Colleges"; John R. Hill, "Presidential Perception: Administrative Problems and the Needs of Public Black Colleges," *JNE* 44, no. 1 (1975): 53–62.
73. Jeffrey Selingo, "Why Minority Recruiting Is Alive and Well in Texas," *CHE* 46, no. 13 (1999): A34–A36.
74. American Association of University Professors Committee on Historically Black Institutions and Scholars of Color, "Historically Black Colleges and Universities: Recent Trends (2007)," (2006) at www.aaup.org (accessed 8/12/2008).
75. Lydia Lum, "Funding Friction," *BIHE* 18, no. 15 (2001): 34–35; "Almost No Science Grants to Black Colleges and Universities: The Mysterious Behavior of the President of Prairie View A&M," *JBHE*, no. 33 (2001): 68–69; Todd Ackerman, "Prairie View Alums Claim Hiring Racist; A&M Regents Blasted for President Search," *Houston Chronicle*, September 30, 1994; Kendra Hamilton, "When the Campus Becomes a Battleground," *BIHE* (2002): 20–24; Nancy T.

King, "The State of Governance at HBCUs" (Paper presented at the Governance Conference, Morris Brown University, October 18, 2002); Armando Villafranca, "Prairie View Board Hires New President; Smithsonian Official Named to Post," *Houston Chronicle*, September 29, 1994.
76. Such programs resonated with federal impulses, ongoing since the Carter administration, to replace institution-based aid with support for individual students. That trend is widely seen as having its roots in conservative distaste for basing federal aid on anything other than individuals' demonstrated potential for achievement along conventional academic standards. See Morris, *Elusive Equality*, 201.
77. American Association of University Professors Committee on Historically Black Institutions and Scholars of Color, "Historically Black Colleges and Universities: Recent Trends (2007)."
78. Forty-nine percent of all Texas college students received their degrees within six years of enrolling; only 27 percent of black students in the state did so. See "Texas' Efforts to Cushion the Ban on Race-Sensitive Admissions," *JBHE*, no. 22 (1998–1999): 54.
79. Roberts, quoted in Hamilton, "Prairie View Has Become Negro Educational and Cultural Capital of Texas"; Matthew Tresaugue, "Prairie View's Past Lost in the Present?" *Houston Chronicle*, July 11, 2005; "Historian James A. Wilson Jr. to lead Prairie View A&M Honors Program," August 26, 2009 at www.pvamu.edu (accessed 9/12/09). In 2000, Texas Instruments gave PVAMU a grant of $1.63 million to develop the school's programs and facilities in electrical engineering. This indicates both the successes of PVAMU engineering faculty in establishing scientific credibility and the powerful late-twentieth century role of private industry in the growth of the land-grant system. Federal contributions to HBCU science and engineering in this period are explored in Chapter 7. See "Notable Minority-Related Grants to Institutions of Higher Education," *JBHE*, no. 29 (2000): 146.
80. Lydia Lum, "Panel Offers Diversity Goals for Texas HBCUs," *BIHE* 17, no. 7 (2000): 15. In 2000, Houston was the fourth largest city in the United States. If Prairie View, roughly an hour's drive away, wished to grow with or without regard to its demographic make-up, Houston would be a significant source of new students.
81. Roy Bragg, "Minority Plan May Cost $1 Million," *Bryan-College Station Eagle*, December 22, 1980.
82. Morris, *Elusive Equality*, 201; Willie and MacLeish, "The Priorities of Presidents of Black Colleges," 144.
83. Lum, "Panel Offers Diversity Goals for Texas HBCUs"; Sara Hebel, "Federal Officials Say Texas Desegregation Plan Needs Revision," *CHE* 47, no. 11 (2000): A32; Patrick Healy, "A Lightning Rod on Civil Rights," *CHE* 46, no. 4 (1999): A42–A44.
84. Healy, "A Lightning Rod on Civil Rights"; Terence J. Pell, "Texas Must Choose between Court Order and Clinton Edict," *Wall Street Journal*, April 2, 1997.
85. "Next Leader up to Challenge of Reviving Prairie View A&M," *Austin American-Statesman*, June 10, 2003, A10.

7. Standards and the "Problem" of Affirmative Action

1. For a summary and critique of this argument, see Joe R. Feagin, Hernan Vera, and Nikitah Imani, *The Agony of Education: Black Students at White Colleges and Universities* (New York: Routledge, 1996), 1-4; and Christopher Newfield, *Unmaking the Public University* (Cambridge, MA: Harvard University Press, 2008), 107-124.
2. Walter R. Allen and Joseph O. Jewell, "A Backward Glance Forward: Past, Present, and Future Perspectives on Historically Black Colleges and Universities," *The Review of Higher Education* 25, no. 3 (2002): 241-261; Avery Gordon, "The Work of Corporate Culture: Diversity Management," *Social Text* 13, no. 3 (1995): 3-30; Claire Jean Kim, "Clinton's Race Initiative: Recasting the American Dilemma," *Polity* 33, no. 2 (2000): 186; Shirley M. Malcom, Daryl E. Chubin, and Jolene K. Jesse, *Standing Our Ground: A Guidebook for STEM Educators in the Post-Michigan Era* (Washington, DC: American Association for the Advancement of Science, 2004), 49; Jeffrey Selingo, "U. of Texas Ends Minority Hiring Plan," *CHE* 45, no. 19 (1999): A38.
3. G. Kemble Bennett, "From the Dean's Desk," *Engineering News* (Winter 2006); Peter Schmidt, "Texas A&M Raises Minority Enrollments without Race-Conscious Admissions," *CHE* 51, no. 21 (2005); Claire Swedberg, "Saluting Our Schools: Texas A&M Reaches out to Restore Diversity," *Diversity/Careers* Winter/Spring (2004-2005); Cass R. Sunstein, "Affirmative Action in Higher Education: Why Grutter Was Correctly Decided," *JBHE* 41 (2003): 80-83.
4. Swedberg, "Saluting Our Schools"; Laura Schreier, "Racism Subtler Than Video, A&M Students Say," *Dallas Morning News*, November 30, 2006. Of the 7,104 freshmen who entered TAMU that year, only 256 were black and 1,001 Hispanic, far short of the level at which these minorities are represented in the state. Paul Burka, "Agent of Change," *Texas Monthly* (2006): 8.
5. Schmidt, "Texas A&M Raises Minority Enrollments"; "Reality Check: Texas Top Ten Percent Plan," *Hispanic Outlook* (2004), cited in Malcom, Chubin, and Jesse, *Standing Our Ground*, 23. On the "plateauing" of minority enrollments, see J. A. Youngman and C. J. Egelhoff, "Best Practices in Recruiting and Persistence of Underrepresented Minorities in Engineering: A 2002 Snapshot" (Paper presented at the ASEE/IEEE Frontiers in Education Conference, Boulder, Colorado, November 5-8, 2003), 14.
6. Schmidt, "Texas A&M Raises Minority Enrollments"; Greg Winter, "Texas A&M Ban on 'Legacies' Fuels Debate on Admissions," *New York Times*, January 13, 2004. On the role of individual traits in minority engineering, see Richard Tapia, Daryl E. Chubin, and Cynthia Lanius, *Promoting National Minority Leadership in Science and Engineering* (Houston, Tx: Rice University/National Science Foundation, 2000), 5; and Heather Dryburgh, "Work Hard, Play Hard: Women and Professionalization in Engineering—Adapting to the Culture," *Gender and Society* 13, no. 5 (1999): 664-682.
7. Jen S. Schoepke, "What Does It Mean to Be White? The Experience of Male Engineers in a Research University" (Research proposal, University of Wisconsin-

Madison, 2007); Patricia Gurin and Edgar Epps, *Black Consciousness, Identity, and Achievement: A Study of Students in Historically Black Colleges* (New York: Wiley, 1975). On the marginality of anti-racist research, see Tania Das Gupta, "Teaching Anti-Racist Research in the Academy," *Teaching Sociology* 31, no. 4 (2003): 456–468.

8. David K. Brown, "The Social Sources of Educational Credentialism: Status Cultures, Labor Markets, and Organizations," *Sociology of Education* 74 (2001): 19–34.
9. Author interview with Kenneth Ashworth, July 17, 2005.
10. Stephanie Pattillo, "Black Former Students Reflect on A&M Life—Past & Present," *Battalion*, September 30, 1993; "'70s Important Time in A&M History," *Battalion*, February 8, 2000; Mexican American Legal Defense and Educational Fund et al., "Blend It, Don't End It: Affirmative Action and the Texas Ten Percent Plan after Grutter and Gratz" (executive summary), *Harvard Latino Law Review* 8 (2004): 34–50.
11. William H. Mobley, "Education Is a Way of Investing in Minorities," *Bryan-College Station Eagle*, February 21, 1992; Author interview with Kenneth Ashworth, July 17, 2005.
12. Author interview with Percy Pierre, July 26, 2005; Lydia Lum, "Panel Offers Diversity Goals for Texas HBCUs," *BIHE* 17, no. 7 (2000): 14–15; Mexican American Legal Defense and Educational Fund et al., "Blend It, Don't End It."
13. Lisa McLoughlin, "The Exceptional Woman Reconsidered: Two Successful Types of Undergraduate Women Engineering Students" (Paper presented at Locating Engineers: Education, Knowledge and Desire, International Network for Engineering Studies, Virginia Tech University, September 9-12, 2006).
14. Burka, "Agent of Change," 7. In 1991, TAMU established the Race and Ethnic Studies Institute, primarily through involvement with faculty in sociology. The unit was intended to provide resources for community outreach, scholarly work, or institutional assessment, but seems not to have consistently covered much ground in any of those as it changed directors and missions over the years. In 2007, the institute was relaunched as an interdisciplinary effort, with a new advisory board, focusing on scholarship. See "Director's Message" at resi.tamu.edu (accessed 1/10/08).
15. These earliest African American enrollees at College Station were attendees of an NSF institute in Earth Science, not matriculated students. Pattillo, "Black Former Students Reflect on A&M Life"; Texas A&M University, "Leroy Sterling: Breaking the Color Barrier, First African-American Students at A&M June 1963" in TAMUA. Note the words chosen for the *Battalion's* 1963 headline: "Three Negroes Enroll Quietly for 1st Term," *Battalion*, June 6, 1963. On comparisons between Gates and Rudder, see Burka, "Agent of Change."
16. Texas A&M University, "In Fulfillment of a Dream: African-Americans at Texas A&M University" (exhibition catalog), ed. J. Angus Martin (2001), in TAMUA; Cassie Patterson, "Flaws Realized: Affirmative Action at Texas A&M University," (typescript, 2004), in TAMUA, 2; John R. Thelin, "Looking for the Lone Star Legacy: Higher Education in Texas," *History of Education Quarterly* 17, no. 2 (1977): 223; "ASC Will Integrate Facilities in Fall," *Dallas Morning News*,

July 11, 1962; Fred King, "Officials Hope for OK on Racial Compliance in Texas Universities," *Houston Post,* January 2, 1981; author interview with Jeanne Rierson, June 5–6, 2002. Although a "Black Culture Advanced and Unified" group appears to have existed earlier, in 1970 African American students at College Station formed the Black Awareness Committee, which lobbied for improved minority recruitment and created a 56-page booklet for this purpose entitled "Power Through Education at Texas A&M." The group faced some hostility through the 1980s. See "A&M Black Students Form Ad Hoc Affairs Committee," *Battalion,* February 21, 1969; "Blacks Form Campus Group," *Battalion,* September 8, 1970; Kathy Brueggen, "Racist Charges Prompt Answers," *Battalion,* November 20, 1973; Jerry Rosiek, "Committee Ignoring Its Educational Role," *Battalion,* November 9, 1987.

17. Patterson, "Flaws Realized," 3; Kim Schmidt, "Vandiver Unveils Minority Plan; Council Approves Faculty Senate," *Battalion,* January 20, 1983; Kara Bounds, "Report: A&M Lags When It Comes to Minority Concerns," *Bryan-College Station Eagle,* May 12, 1992; Christine Mallon, "Minority Funding Boosted," *Battalion,* April 4, 1984,
18. Lawrence E. Young, "A&M Luring, Keeping More Minorities," *Dallas Morning News,* December 28, 1987; Christine Mallon, "Minority Funding Boosted"; Patterson, "Flaws Realized," 3.
19. "A&M Helps Minority Students Realize Medical School Dreams," *Bryan-College Station Eagle,* February 10, 1992; "A&M College of Medicine Focuses on Minority Recruitment," *Battalion,* February 18, 1992; John R. Thelin, *A History of American Higher Education* (Baltimore: Johns Hopkins University Press, 2004), 334.
20. Author interview with Richard Perry, June 5, 2002; C. W. Crawford, *One Hundred Years of Engineering at Texas A&M, 1876–1976* (1976), 76.
21. "Carl A. Erdman" (obituary), *Bryan-College Station Eagle,* June 15, 1995.
22. One newspaper article cites an ASEE finding that TAMU's engineering departments led the nation in awarding undergraduate degrees to minority students, granting 115 to black and Hispanic students in 1986, but I have not found ASEE publications making this point. Lawrence E. Young, "A&M Luring, Keeping More Minorities."
23. Karan L. Watson and Mary R. Anderson-Rowland, "Interfaces between the Foundation Coalition Integrated Curriculum and Programs for Honors, Minority, Women, and Transfer Students," in *ASEE/IEEE Frontiers in Education* (1995).
24. Mackenzie Garfield, "Texas A&M Co-Founder Receives Formal Recognition," *Battalion,* February 9, 2007.
25. Urban Institute, "Evaluation of the National Science Foundation Louis Stokes Alliances for Minority Participation Program" at www.urban.org (accessed 9/22/07).
26. Author interviews with Jeanne Rierson and Karan Watson, June 5–6, 2005.
27. Dryburgh, "Work Hard, Play Hard"; Urban Institute, "Evaluation of the National Science Foundation," 35.
28. Juan C. Lucena, *Defending the Nation: U.S. Policymaking to Create Scientists and Engineers from Sputnik to the 'War against Terrorism'* (Lanham, MD: University Press of America, 2005), 140–147.

29. H. Roberts Coward, Catherine P. Ailes, and Roland Bardon, *Progress of the Engineering Education Coalitions* (Arlington, VA: Engineering Education and Centers Division, National Science Foundation, 2000), 5–7.
30. Ibid., 21, 46.
31. Thelin, *A History of American Higher Education*, 278–279; Roald F. Campbell and William M. Boyd II, "Federal Support for Higher Education: Elitism Versus Egalitarianism," *Theory into Practice* 9, no. 4 (1970): 232–238.
32. Coward, Ailes, and Bardon, *Progress of the Engineering Education Coalitions*, 17–19.
33. Scot Walker, "A&M Gets Grant to Improve Curriculum for Engineers," *Battalion*, October 6, 1988; "Erdman is Coalition Director," *Bryan-College Station Eagle*, January 17, 1994; Carl Erdman, "The Foundation Coalition," in *ASEE/IEEE Frontiers in Education* (1994): 646; Jeffrey Froyd, Debra Penberthy, and Karan L. Watson, "Good Educational Experiments Are Not Necessarily Good Change Processes" (Paper presented at the ASEE/IEEE Frontiers in Education, Kansas City, MO, 2000), 4; Watson and Anderson-Rowland, "Interfaces." When the NSF extended funding for the Foundation Coalition for years six through ten, the original schools were joined by the University of Wisconsin at Madison and the University of Massachusetts at Dartmouth.
34. Erdman, "The Foundation Coalition," 645; David Cordes, "NSF Foundation Coalition: The First Five Years," *Journal of Engineering Education* 88, no. 1 (January 1999): 73–77.
35. Erdman, "The Foundation Coalition," 646.
36. Ibid.
37. Watson and Anderson-Rowland, "Interfaces"; Victor Wilson, Taanya Monogue, and Cesar O. Malave, "First Year Comparative Evaluation of the Texas A&M Freshman Integrated Engineering Program," in *ASEE/IEEE Frontiers in Education* (1995).
38. Agyeman Boateng, "Viceroys of Difference: Minority Engineering Program Directors at Mid-Atlantic Universities" (Master's Thesis, Drexel University, 2006); Martha Heath Bowman and Frances K. Stage, "Personalizing the Goals of Undergraduate Research," *Journal of College Science Teaching* 32, no. 2 (1999).
39. Erdman, "The Foundation Coalition," 647; Coward, Ailes, and Bardon, "Progress of the Engineering Education Coalitions," 16; Wilson, Monogue, and Malave, "First Year Comparative Evaluation." Coward later finds that "student networks were well informed and highly judgmental" about different instructors (p. 23).
40. Coward, Ailes, and Bardon, "Progress of the Engineering Education Coalitions"; Watson and Anderson-Rowland, "Interfaces"; Wilson, Monogue, and Malave, "First Year Comparative Evaluation."
41. Coward, Ailes, and Bardon, "Progress of the Engineering Education Coalitions," 34; Watson and Anderson-Rowland, "Interfaces."
42. Coward, Ailes, and Bardon, "Progress of the Engineering Education Coalitions," 21, 40, 42; Froyd, Penberthy, and Watson, "Good Educational Experiments Are Not Necessarily Good Change Processes", 4; Watson and Anderson-Rowland, "Interfaces."

43. Coward, Ailes, and Bardon, "Progress of the Engineering Education Coalitions," 21.
44. In an overview of curricular change models, Watson herself noted the importance of rewards in an analysis of institutional change processes. Froyd, Penberthy, and Watson, "Good Educational Experiments Are Not Necessarily Good Change Processes", 2; Tapia, Chubin, and Lanius, "Promoting National Minority Leadership in Science and Engineering," 6–8.
45. "Minority Education Target of Inquiry," *San Antonian Express*, January 27, 1993; Lydia Lum, "Texas A&M to Leave Race out of Admissions Decisions," *CHE* 20, no. 23 (2004) 10; Texas A&M University, "In Fulfillment of a Dream."
46. Jeffrey Browne, quoted in Robin Roach, "Despite Growth, Minority Numbers Still Low at A&M," *Battalion*, June 8, 1993.
47. W. G. Bowen and D. Bok, *The Shape of the River: Long-Term Consequences of Considering Race in College and University Admissions* (Princeton, NJ: Princeton University Press, 1998).
48. Beth Panitz, "Policy Shifts," *ASEE Prism* 6 (1996): 15–16; Joshua Rolnick, "Minority Enrollment Drops in Graduate Science Programs," *CHE* 45, no. 4 (1998): A48; Jeffrey Selingo, "Why Minority Recruiting Is Alive and Well in Texas," *CHE* 46, no. 13 (1999): A34–A36; Texas Louis Stokes Alliance for Minority Participation, "Findings: Student Outcomes-Graduation" [report, December 10, 2002], at www.tamu.edu (accessed 6/21/07).
49. Karin Fischer, "Class-Rank Plan Faces Trouble in Texas," *CHE* 51, no. 33 (2005): A25; Peter Schmidt, "A New Route to Racial Diversity," *CHE* 51, no. 21 (2005): A22; Marta Tienda et al., *Closing the Gap? Admissions and Enrollments at the Texas Public Flagships before and after Affirmative Action* (Woodrow Wilson School of Public and International Affairs, Princeton University, 2003). For an overview of recent legislative attention to Texas' Top 10 percent law, see "10% Admissions—The Full Impact," *Inside Higher Education*, April 6, 2009, at www.insidehighered.com (accessed June 29, 2009).
50. Tienda et al., *Closing the Gap?*, 16; Mexican American Legal Defense and Educational Fund et al., "Blend It, Don't End It," 45.
51. James M. Graham, Rita Caso, and Jeanne Rierson, "The Effect of the Texas A&M University System Amp on the Success of Minority Undergraduates in Engineering: A Multiple-Outcome Analysis" (Paper presented at the American Society for Engineering Education Annual Conference and Exposition, 2001); Selingo, "U. of Texas Ends Minority Hiring Plan."
52. Claire Jean Kim points out that the "Democratic party platform of 1992 . . . was the first in almost 30 years to make no mention of redressing racial injustice," in "Clinton's Race Initiative: Recasting the American Dilemma," 189; Lum, "Texas A&M to Leave Race out of Admissions Decisions."
53. Bennett, "From the Dean's Desk."
54. On Gates's failure to correct long-standing patterns of minority exclusion at College Station, see Royce West, quoted in Lum, "Texas A&M to Leave Race out of Admissions Decisions," 10.
55. More than 300 white students were admitted as legacy cases in 2002 and 2003, which is "nearly as many as the total number of black students admitted

to the university in those years." Winter, "Texas A&M Ban on 'Legacies' Fuels Debate on Admissions."

56. Kim, "Clinton's Race Initiative"; Newfield, *Unmaking the Public University.*
57. Burka, "Agent of Change"; Kim, "Clinton's Race Initiative"; Lum, "Texas A&M to Leave Race out of Admissions Decisions"; Robert M. Gates, "Statement by Texas A&M President Robert M. Gates, December 3, 2003" (2003), at www.tamu.edu (accessed 6/22/07).
58. Author interview with Jeanne Rierson, June 5–6, 2002.
59. Youngman and Egelhoff, "Best Practices in Recruiting and Persistence of Underrepresented Minorities in Engineering," 16.
60. Burka, "Agent of Change."
61. U.S. Department of Education, "White House Initiative on Historically Black Colleges and Universities: A Brief History" at www.ed.gov (accessed 9/24/07).
62. Philip J. Sakimoto and Jeffrey D. Rosendahl, "Obliterating Myths about Minority Institutions," *Physics Today* 58, no. 9 (2005).
63. NASA's predecessor agency, the National Advisory Committee on Aeronautics, can be seen as one of many attempts to "Create channels of funding between the government and universities." Robert Kohler, in Robert Kohler, Ellis Hawley, and Nathan Reingold, "Government Science," *Isis* 78, no. 4 (1987): 582; Thelin, *A History of Higher Education,* 335.
64. U.S. Department of Education, "White House Initiative on Historically Black Colleges and Universities" at www.ed.gov (accessed 9/22/2007); Froyd, Penberthy, and Watson, "Good Educational Experiments Are Not Necessarily Good Change Processes"; Robert Bruce Slater, "Rating the Science Departments of Black Colleges and Universities," JBHE, no. 4 (1994): 90–96; Also, author interview with Richard Wilkins, August 4, 2005. For the full text of President Carter's Executive Order, No. 12232, see www.presidency.ucsb.edu (accessed 10/1/2007).
65. Bowen and Bok historicize these arguments, as do many policy documents on minority education. See, for example, Malcom, Chubin, and Jesse, *Standing Our Ground;* Mfanya Donald Tryman, *A Study of the Minority College Programs at the NASA Johnson Space Center* (Houston, TX: NASA, 1987); Bowen and Bok, *The Shape of the River.*
66. Most prominent among sources for this literature is the Center for Equal Opportunity. See, for example, Robert Lerner and Althea K. Nagai, *Pervasive Preferences: Racial and Ethnic Discrimination in Undergraduate Admissions across the Nation* (Washington, DC: Center for Equal Opportunity, 2001). For discussion of anti-affirmative action arguments, see Newfield, *Unmaking the Public University,* 92–102. Peter Schmidt, "From 'Minority' to 'Diversity'," *CHE* 52, no. 22 (2006); Ronald Roach, "Affirmative Action Fallout: Graduate-Level Programs Once Aimed at Minorities Now Opening Up to All Students in Effort to Avoid Legal Challenges," *BIHE* 22, no. 11 (2005): 28–29.
67. American Association of University Professors Committee on Historically Black Institutions and Scholars of Color, "Historically Black Colleges and Universities: Recent Trends (2007)," (2006) at www.aaup.org (accessed 8/12/2008). See also "Almost No Science Grants to Black Colleges and Universities: The

Mysterious Behavior of the President of Prairie View A&M," *JBHE*, no. 33 (2001): 68–69; "Do the Right Thing and Get Rid of Hines" (editorial) *University Faculty Voice* (2002), at www.facultyvoice.com (accessed 10/19/2007); Kendra Hamilton, "When the Campus Becomes a Battleground," *BIHE* (2002); Dale Lezon, "Hines Given $400,000 to Quit PV," *Houston Chronicle*, May 30, 2002.
68. PVAMU, College of Engineering, "Report of the College of Engineering for the Prairie View A&M University Long-Range Development Plan 1981–1987" (Prairie View, TX: Prairie View A&M University, 1980) in PVAMUA.
69. Leo J. Klosterman, Lloyd S. Swenson Jr., and Sylvia Rose, eds., *100 Years of Science and Technology in Texas: A Sigma Xi Centennial Volume* (Houston, TX: Rice University Press, 1986).
70. Amilcar Shabazz, *Advancing Democracy: African Americans and the Struggle for Access and Equity in Higher Education in Texas* (Chapel Hill: University of North Carolina Press, 2004); author interview with Kenneth Ashford, July 17, 2005.
71. Author correspondence with Thomas Fogerty, December 19, 2005; David E. Sanger, "The Troubled Faculty: Corporate Teaching Help Drops," *New York Times*, August 18, 1985.
72. Roy Kleinsasser, "Legislator Disputes Prairie View Funds," *Bryan-College Station Eagle*, January 2, 1979.
73. Author correspondence with Thomas Fogerty; NASA, "Fiscal Year 1998 Annual Performance Report to the White House Initiative Office on Historically Black Colleges and Universities," at www.nasa.gov (accessed 11/30/08); Lewis Harris, Phillip Hopkins, and Sue A. Krenek, "Computers," in *100 Years of Science and Technology in Texas*, 220; Mfanya Donald Tryman, *Organization Development at the Johnson Space Center: A Case Study of Affirmative Action* (Notre Dame, Indiana: Marquette Monographs, 1989).
74. NASA, "Fiscal Year 1998 Annual Performance Report to the White House Initiative Office on Historically Black Colleges and Universities," 21. Author interview with Richard Wilkins, August 4, 2005; author correspondence with Thomas Fogerty. As the 1990s began, Prairie View still had no administrative office overseeing sponsored research, a partial cause of NASA's rejection of Prairie View's proposal to become one of five HBCU URCs supported by the agency. But JSC offered a great deal of guidance to the school and Prairie View succeeded in the next round of URC funding, receiving that designation in 1995.
75. NASA, "NASA Awards PVAMU $1 Million-Plus Grant" (press release, August 31, 2004) at www.pvamu.edu (accessed 11/11/07).
76. Ibid.; Texas Engineering Experiment Station, "New Institute Encourages Students to Pursue Space Industry Careers" at engrportal.tamu.edu (accessed 11/22/08).
77. NASA, "Minority University Research and Education Program: Science, Aeronautics and Technology Fiscal Year 2000 Estimates Budget Summary" (2000), at www.nasa.gov (accessed 11/11/07); U.S. Department of Education, "White House Initiative"; Charles Dervarics, "Dialing for Dollars: Against Heavy Competition, HBCUs Need Savvy, Expertise to Win Department of Defense Funding," *BIHE* 14, no. 13 (1997): 22.

78. "Network Resource and Training Sites" at muspin.gsfc.NASA.gov (accessed 10/10/07).
79. Ibid.
80. Dervarics, "Dialing for Dollars," 22.
81. Author interview with Richard Wilkins, August 4, 2005. A similar set of programs emerged under NASA's Minority University and College Education and Research Partnership Initiative (MUCERPI), which addressed both curricular and research development at minority institutions. NASA, "Minority University Research and Education Program"; Sakimoto and Rosendahl, "Obliterating Myths about Minority Institutions."
82. The anxiety of science and engineering faculty about working with minority graduate students is reflected in Clarence G. Williams, *Technology and the Dream: Reflections on the Black Experience at MIT, 1941–1999* (Cambridge, MA: MIT Press, 2001).
83. Author interview with Richard Wilkins, August 4, 2005; author correspondence with Thomas Fogarty. Dervarics, "Dialing for Dollars."
84. Jennifer S. Light, *From Warfare to Welfare: Defense Intellectuals and Urban Problems in Cold War America* (Baltimore: Johns Hopkins University Press, 2005), 1–3.
85. Quoted in Roger D. Launius, "Compelling Rationales for Spaceflight? History and the Search for Relevance," in *Critical Issues in the History of Spaceflight*, ed. Steven J. Dick and Roger D. Launius (Washington, DC: NASA, 2006), 51, 53.
86. Ibid., 50.
87. Malcom, Chubin, and Jesse, *Standing Our Ground*, 47–48.
88. NASA, "Fiscal Year 1998 Annual Performance"; NASA, "Minority University Research and Education Program."
89. "$5 Million NASA Grant Assures Prairie View A&M Leadership in Space Radiation Research" at www.pvamu.edu (accessed 11/22/08).
90. Brown, "The Social Sources of Educational Credentialism," 26.
91. Steven Shapin, *A Social History of Truth: Civility and Science in Seventeenth-Century England* (Chicago: University of Chicago Press, 1994), 415.

8. Conclusion

1. An unusual and very helpful resource here is Alice L. Pawley, "Gendered Boundaries: Using a 'Boundary' Metaphor to Understand Faculty Members' Descriptions of Engineering" (Paper presented at ASEE/IEEE Frontiers in Education, Milwaukee, WI, 2007).
2. Mark T. Holtzapple and W. Dan Reece, *Foundations of Engineering*, 2d ed. (Boston: McGraw-Hill, 2003).
3. Ibid., 24–31, 44.
4. Elaine Seymour and Nancy Hewitt, *Talking about Leaving* (Boulder, CO: Westview Press, 1997).
5. Holtzapple and Reece, *Foundations of Engineering*, 26.
6. Shapin offers a model for tracing how trust and identity in scientific communities intersect. Steven Shapin, *A Social History of Truth: Civility and Science in*

Seventeenth-Century England (Chicago: University of Chicago Press, 1994). Christopher Newfield compellingly outlines how prevailing definitions of diversity most recently have foreclosed understanding of the racial advantage and disadvantage in U.S. higher education; *Unmaking the Public University* (Cambridge: Harvard University Press, 2008), 92–110.

7. John Brooks Slaughter, "The 'New' American Dilemma," in NACME, *Confronting the 'New' American Dilemma: Underrepresented Minorities in Engineering: A Data-Based Look at Diversity* (White Plains, NY: NACME, 2008).
8. Claire Jean Kim, "Clinton's Race Initiative: Recasting the American Dilemma," *Polity* 33, no. 2 (2000): 196.
9. Arthur T. Johnson and Rosemary L. Parker, "The Effects of Technology on Diversity, or When Is Diversity Not Diversity?" (Paper presented at the American Society for Engineering Education Annual Conference and Exposition, 2001).
10. Bryan McKinley Jones Brayboy, "The Implementation of Diversity in Predominantly White Colleges and Universities," *JBS* 34, no. 1 (2003): 72–86.
11. Shirley Ann Jackson, "The Perfect Storm: A Weather Forecast," speech to the annual meeting of the American Association for the Advancement of Science (February 14, 2004), available at www.rpi.edu (accessed 11/20/08).
12. Johnson and Parker, "The Effects of Technology on Diversity, or When Is Diversity Not Diversity?"
13. Pawley, "Gendered Boundaries."
14. Daryl E. Chubin, "Transcending the Places That Hold Us: Public Policy and Participation in Science" (Paper presented at Workshop 2000: A Joint Conference of the American Association for the Advancement of Science, the Emerge Alliance, and the National Science Foundation, Atlanta, GA, February 24, 2000).
15. Brayboy, "The Implementation of Diversity in Predominantly White Colleges and Universities."
16. William Wulf, cited in Foundation Coalition, "Women and Minorities in Engineering" at www.foundationcoalition.org (accessed 3/5/2005).
17. Mark Evans, "New Institute Encourages Students to Pursue Space Industry Careers" (press release, Texas Engineering Experiment Station, 2003).
18. On continuing associations of economic opportunity and race, see Fred R. Harris and Lynn A. Curtis, *Locked in the Poorhouse* (New York: Roman and Littlefield, 1998); "Income Gap between Families Grows," *New York Times*, November 13, 2007.
19. Jonson William Miller, "Citizen Soldiers and Professional Engineers: The Antebellum Engineering Culture of the Virginia Military Institute" (Ph.D. diss., Virginia Polytechnic and State University, 2008); Heather Dryburgh, "Work Hard, Play Hard: Women and Professionalization in Engineering—Adapting to the Culture," *Gender and Society* 13, no. 5 (1999): 664–682.
20. Pawley, "Gendered Boundaries."
21. Peter Wood, "How Our Culture Keeps Students Out of Science," *CHE* 54, no. 48 (2008): A56.
22. Kim, "Clinton's Race Initiative": 197.

23. Upon replacing George W. Bush, Barack Obama retained Robert Gates as U.S. Secretary of Defense. Obama's perceptions of Gates's approach to diversity in higher education, and Obama's understanding of how ideologies regarding race might intersect with defense operations are beyond the scope of this book, but offer intriguing areas for further study.
24. V. P. Franklin, "Hidden in Plain View: African American Women, Radical Feminism, and the Origins of Women's Studies Programs, 1967–1974," *Journal of African American History* 87 (2002): 433–435; William H. Watkins, *The White Architects of Black Education: Ideology and Power in America, 1865–1954* (New York: Teachers College Press, 2001).

Index

Ability, individual, 49, 77, 101, 144, 190, 223–224n14, 225n5, 235n63; and racial beliefs, 45, 67, 69, 145. *See also* Color blindness; Merit

Accountability, 101, 120, 129, 135, 184

Accreditation, 5, 32, 33, 58, 61, 98, 101–102, 104, 106, 111, 126, 130, 140, 152, 181, 182, 211, 222n4, 242n72. *See also* Accreditation Board for Engineering and Technology

Accreditation Board for Engineering and Technology (ABET), 5, 102, 106, 181, 182, 211

Activism, 14, 109, 163, 191, 234n54; race activism, 8, 9, 12, 13, 14, 19, 29, 38, 49, 59, 64, 67, 100, 111, 113, 132, 147, 149, 151, 157, 162, 164, 177, 232n28, 251n8. *See also* Civil rights movement; Progressivism; Racial conflict

Admissions, 45, 51, 77, 110, 112, 125, 139, 177–178, 180, 204; criteria for, 4, 5, 9, 10, 11, 104–109, 113, 116, 119, 126, 147, 161, 170, 218; inclusive, 109, 147, 170, 175; and income, 178, 189, 242n67; legacy admissions, 190, 265–266n54; open, 26, 27, 72, 104–110, 159; "qualified" vs. "qualifiable," 98, 100, 165, 210; selective, 5, 27, 88, 94, 98–100, 108, 109, 113, 116, 126, 147, 161, 165, 222n3. *See also* Affirmative action; Color blindness; Eligibility; Integration

Advising, student, 99, 105, 124

Affirmative action, 1, 3, 7, 82, 96, 103, 112, 143–144, 193, 202–203, 222n3; laws, 14, 141, 145, 146, 167; opposition, 3, 7, 14, 144, 170, 183, 187–189, 207, 211, 216, 251n3; Texas A&M, 171–173, 177, 183, 187–190, 202. *See also specific court cases*; Color blindness; Reverse discrimination; Supreme Court

Agriculture, 6, 12, 90, 128; agricultural experiment stations, 25, 41; extension programs, 41, 61; research, 25, 40, 61, 194–195; and segregation, 37–43, 62–65; training, 20, 21, 23, 25, 27, 29, 31, 36–43, 51, 62, 68, 148, 149, 158, 194, 233n39. *See also* Rural life

Alabama Poytechnic Institute, 44

Alcorn State University, 41

Alfred P. Sloan Foundation, 95, 154, 159

Alger, Horatio, 163, 173

Alliance for Minority Participation, 174. *See also* Louis Stokes Alliance for Minority Participation

Alta Vista Agricultural College, 148

American Academy for the Advancement of Science, 124, 137
American Council on Education, 95, 118–119, 131, 135, 246n17
American Society for Engineering Education (ASEE), 93, 138, 211, 262n22
Amherst College, 43
Anderson, Alan, 92
Anderson, James D., 18
Architectural programs, 5, 22, 68, 132, 151, 154, 156, 161
Armour Institute of Technology, 83–86
Armour Research Foundation, 85, 121. See also Illinois Institute of Technology Research Institute
Ashworth, Kenneth, 152, 175, 257n58
Asians, 2, 6, 173, 180
Atomic Energy Commission, 53, 141, 151, 231n15
Attainment: academic, 14, 24, 60, 68, 179, 208, 223n14; career, 1, 41, 61, 67–68, 71, 153, 164, 173–174, 206; economic, 17, 22. See also Enrollment patterns; Graduates, engineering; Mobility

Bakke case. See Regents of the University of California v. Bakke
Baltimore, 12, 20–21, 23, 24, 26–30, 35–36, 38–39, 42, 43, 51, 54, 56–57, 59, 63, 72–74
Baltimore Sun, 33, 57
Bay Area Rapid Transit (BART), 119–120, 133, 134
Becton, Julius, 166, 169
Behavior, social, 22, 39, 46, 57, 67, 68, 69–71, 88–89, 96, 134, 144, 147, 184, 241n60
Bennett, W. Kemble, 189
Biology (and race), 12, 68, 70, 96, 224n16. See also Racism
Black consciousness, 88–89, 162–164, 177, 191, 207. See also under Identity
Bluestone, Daniel, 84, 87
Bolden, Charles, 16
Bowie State University, 29, 61
Bradley, Omar, 68
Brayboy, Bryan, 210, 212
Bridge programs, 180, 183
Brown, David K., 203
Brown v. Board of Education of Topeka, 12, 16, 21, 66, 72, 76
Bugliarello, George, 91, 105, 110, 116, 122–123, 125, 126, 129–131, 132, 137–138, 140, 242n72, 248n45

Bullock, Henry Allen, 150
Bush, George H. W., 14
Bush, George W., 14, 169, 172–173, 188, 269n23
Bush, Vannevar, 48
Byrd, Harry ("Curly"), 5, 12, 21–22, 27, 29–42, 45–47, 52, 53, 68, 70–71, 226n6, 227n19, 231n22, 232n24, 233n32, 236n78; and College Park, 29, 30–34, 36, 42–43, 50–52, 54–55, 58; criticisms of, 55–57; Eastern Shore development, 50–52, 55–57, 60, 233n31; and integration, 43–45, 59, 64, 70–76, 228n31, 235n69; and John T. Williams, 57–60; views on agriculture and race, 36–42, 63–64

Caldwell, Harmon, 44
Calhoun, J.W., 44
California, 150
California Institute of Technology, 54, 83
Caliver, Ambrose, 65–66, 68–71
Callcott, George, 27, 30, 41, 54, 76
Campus design, 85–87, 91, 238n20
Cantu, Norma, 169
Capitalism, 8, 142, 146, 163, 213–214, 218, 224n19, 230n2, 247n29. See also Globalization
Carter, Jimmy, 157, 192, 256n76
Carver, George Washington, 41
Centenary Bible Institute of Baltimore, 23
Chapman, Oscar, 22, 37, 42, 45–47, 67, 68, 70
Character, 27, 69, 89, 101, 190, 208
Charter schools, 14
Chemical engineering, 33, 42, 123, 150, 195
Chiarulli, Peter, 97
Chicago, 13, 14, 79–93, 103–109, 111, 114–116, 118, 121–124, 127–129, 133, 135–136, 141, 237n5, 238n16, 247n21; civic leaders, 80, 82, 84, 86, 88–89, 92, 108, 109, 115, 118, 121, 136; electoral politics, 92, 109; housing policy, 80, 82–85, 89–92, 114; 116, 121, 127, 128, 130, 133; Loop, 80–81, 83–84, 87, 90–91; planning, 82, 84–85, 87, 91, 103–104, 108, 116, 125, 127–128, 133; South Side, 83–85, 88, 94; suburbs, 80–81, 84, 90, 92, 121
Chicago Committee on Urban Opportunity, 109
Chicago Tribune, 86
Chung, Paul, 106, 107, 126

Civil engineering, 29, 32, 33, 64, 121, 125, 126, 130, 150, 152, 153, 161
Civil Rights Act (1964), 2; Title VI, 157, 168, 175, 189
Civil rights movement, 1, 2, 6, 7, 8, 12, 13, 16, 19, 21–22, 45, 49–50, 73, 77, 82, 87, 90, 92, 96, 111, 119, 142, 143–145, 149, 151, 157, 162, 169–170, 171, 177, 189, 190, 192, 200, 232n28, 250–251n3, 251n8. *See also* Activism
Civil War, 27
Class, social, 8, 13, 19, 38, 46, 49, 77, 81, 83, 90, 91, 93, 105, 108, 133, 140, 146, 163, 209, 215, 225n5, 242n69
Clinton, Bill, 189, 209
Coalitions, research, 181–187, 191, 199, 201–202, 209, 212, 263n33
Cohen, Aaron, 195
Cold War, 7, 14, 87, 121, 136, 192, 230n2
College Park. *See* University of Maryland, College Park
Colonialism, 62, 68, 234n58
Color blindness, 15, 141, 142, 144–146, 171, 176, 179, 190, 191, 251n9. *See also* Ability, individual; Affirmative action; Merit
Columbia University, 65
Community colleges, 102, 155, 180, 183
Commuter students, 84, 90, 91, 93, 132, 162, 257n60
Compton, Arthur, 48
Compton, Karl, 32
Congress, U.S., 4, 59, 95, 243n85
Connolly, John, 152
Conservatism, 4, 5, 11, 14–15, 16, 21, 36, 58, 64, 78, 79, 90, 92, 116, 119, 139–142, 144–145, 152, 162, 169, 187–188, 192, 204, 209, 218, 238n20, 259n76
Co-op programs, 97, 99, 161, 254n31. *See also* Minority engineering programs
Coppin State University, 29
Cornell University, 31, 95, 182, 233n39
Culture, 4, 5, 13, 15, 17, 27, 31, 37, 64, 67, 86–87, 101, 109, 115, 135–136, 158, 171, 191, 197, 200–201, 205, 207, 212, 214, 218, 238n20, 242n69; academic, 101, 106, 131, 137–138, 175–177, 186, 190–191, 211; African American, 3, 24, 67–68, 70–71, 96, 145, 164; corporate, 8, 147; engineering, 8, 111, 131, 135–138, 202, 224n19, 250n68; southern, 38–39, 41, 44; U.S., 10, 45, 48, 101, 115, 210; Western, 14, 133, 223n14. *See also* Behavior, social; Diversity (as goal); Identity; Multiculturalism
Curriculum, academic: curricular reform, 62, 165, 181–191, 242n72; duplication of, 56, 68, 130, 152, 176; extended, 102, 141; remedial courses, 97–99, 101–102, 106, 110, 113, 125, 141, 199, 208, 217. *See also* Interdisciplinary programs; Pedagogy; *specific institutions*

Daley, Richard J., 88, 90–92, 109
Defense: defense industry, 13, 16, 53, 54, 87, 111, 114–117, 121, 124, 128, 135, 140, 141, 147, 163, 196; Department of, 15, 121, 124, 172, 174, 200, 269n23; funding, 16, 115, 121, 124, 174, 192, 198; research, 16, 114–118, 120, 121, 124, 128, 130, 134, 141, 194, 196, 200. *See also* Military, U.S.
Delco, Wilhelmina, 158
Democracy, 8, 19, 43, 140, 189; and science, 48–49, 199
Department of Defense, 121, 124, 192, 198
Department of Education, 23, 157. *See also* Office of Education
Department of Energy, 192
Department of Health, Education and Welfare, 97
Department of Health and Human Services, 192
Department of Housing and Urban Development, 118
Department of Labor, 103, 118
Desegregation. *See* Integration
Detroit, 79, 118, 161, 182
Development, economic, 12, 19–20, 25–26, 27, 30, 32, 38, 41, 45, 49, 82, 87, 119, 121, 145, 149–150, 200, 253n18; role of higher education, 25, 34, 77, 80, 93, 105, 114–122, 127, 134–136, 140, 145, 147, 163, 189, 194, 234n47; urban development, 3, 38, 80–95, 103, 109, 114–134, 200, 238n16, 248n38. *See also* Industrialization; Modernization
Discrimination, 3, 6, 8, 13, 16–17, 20, 21–22, 45–46, 49, 51, 61–62, 79, 80, 82, 88–89, 98, 103, 110, 157, 165, 169, 173, 187, 191, 199, 214, 244n97; vs. "competition," 143–145; structural, 17, 52, 67, 70, 71, 96, 98, 110, 132, 144, 165, 173–174, 191, 197, 198, 204, 211–212, 215. *See also* Reverse discrimination

Diversity (as goal), 2, 4, 6–9, 67, 95–96, 109, 111, 113, 115–116, 139, 142, 144–147, 189, 195, 201, 205–218, 245n1; disincentives, 101–103; IIT, 96, 101–103, 111, 116; and labor, 7, 49, 62, 110, 111, 142, 145–146, 204, 217, 224n20; meaning of, 17, 204; Texas A&M, 15, 144–147, 164, 168–181, 189, 202–203; UIC, 79, 111, 116; University of Maryland, 20, 76. *See also* Integration

Doner, Dean B., 105, 138

Donors. *See* Funding

Dormatories, 35, 53, 74, 86, 91, 156, 160, 175, 177

Dryburgh, Heather, 216

Dual educational systems, 10, 21, 28, 30, 51, 59, 61, 76, 168

Du Bois, W. E. B., 24, 66

Economic development. *See* Development, economic

Education policy. *See* Office of Civil Rights; Office of Education; *and under individual states*

Ehrenreich, Barbara and John, 140

Eisenhower, Dwight D., 94, 135

Electrical engineering, 32, 33, 37, 42, 52, 74, 151, 158, 159, 160–161, 195, 196, 254n28, 259n79

Eligibility: academic, 9–11, 12, 13, 14, 17, 26, 31, 34, 41, 53, 88, 93, 99, 110, 111, 165, 172, 177, 180, 189, 202, 210, 214, 216; funding eligibility, 107, 139–140, 155, 158, 174, 195, 197, 199, 202, 212. *See also* Enrollment patterns; Merit

Employment, 2, 6, 21, 23, 26, 33–34, 36, 40–42, 49, 61, 62, 64, 72, 88, 94–95, 98, 102–105, 108–111, 117, 125, 128, 130, 140, 145–146, 147, 149, 154, 161, 162, 164, 166, 175, 194, 199, 206, 212–213, 216, 243n85, 244n97, 245n3, 249n64, 252n9. *See also* Graduates, engineering; Mobility

Engineering: criticisms of, 114, 118, 124–126, 132, 135–139; objectivity, 13, 65, 111, 131, 152; quantification, 131–134, 137; separation from humanities, 132, 136–137, 177; separation from social thought, 131–139, 190, 215–217, 248n45; social engineering, 114–131, 137, 215–217. *See also* Engineering technology; *individual subfields*

Engineering Education Coalitions, 174, 181–187, 212

Engineering Experiment Stations, 33, 140, 183, 194

Engineering Manpower Commission, 111, 140

Engineering technology, 102, 104, 158, 161, 194

Engineers' Council for Professional Development, 5, 32, 106. *See also* Accreditation Board for Engineering and Technology

Enrollment patterns, 2, 10, 25, 28, 30–31, 54, 100, 105, 135, 145–146, 169, 188, 205, 257n57, 260n5; Eastern Shore, 56, 58, 62; graduate education, 72, 145, 236n83, 256–257n51; IIT, 87, 99; nonminorities at HBCUs, 162, 168, 176, 191; PVAMU, 161–162, 165–166, 168, 176, 255n36; TAMU, 172–173, 175, 177–178, 180, 186–189, 203; UIC, 91, 110. *See also* Eligibility

Environmental issues, 83, 114, 117–120, 124, 125, 127, 128, 129, 135, 138, 206, 215

Epistemology, 12, 131–132, 134, 137, 170, 198, 200, 201

Epps, Edgar, 164, 244n97

Equity, economic, 81–82, 94, 125, 171, 173, 189

Erdman, Carl, 178–179, 183–184, 191

Ethnicity, 6, 11, 39, 63, 64, 92, 95, 101, 145, 147, 164, 171–176, 179, 180, 181, 187, 205, 206, 207, 211, 214, 261n14. *See also* Identity

Extension services, 24, 41, 61–62, 92

Facilities, physical. *See under individual institutions*

Faculty. *See* Retention; Reward systems (academic); Tenure and promotion; *and under individual institutions*

Family, 168, 173, 179, 190, 213; African American, 6, 46, 67, 68, 96, 187; income, 178, 242n67

Financial aid, 9, 104, 156, 169, 172, 188, 259n76. *See also* Scholarships

Fitzgerald, William L., 39–40, 46

Fogelson, Robert, 80

Fogerty, Thomas, 196, 201

Ford Foundation, 114, 131

Foundation Coalition, 181–187, 191, 199, 201–202, 209, 212, 263n33

Franklin, V. P., 218
Funding: Department of Defense, 16, 115, 121, 124, 174, 192, 198; eligibility criteria, 107, 139–140, 155, 158, 174, 195, 197, 199, 202, 212; federal, 4, 21, 23–24, 59–60, 66, 86, 90, 103, 118, 121, 124, 148, 157, 169, 174, 179–182, 186–187, 192–193, 195–203, 212, 225n3, 263n33, 266n74; Illinois, 86, 87, 89, 90, 92, 97, 108; industry, 4, 8, 95, 100, 103, 121, 146, 154, 194, 202, 213; Maryland, 28–29, 32, 33, 35–36, 41–42, 43, 45, 50, 54–61, 66, 75, 76, 232n26, 233n31, 233n39; philanthropic, 24, 41, 83, 95, 103, 114, 121, 131, 154, 181, 228n39; Texas, 144, 151–157, 175, 188, 194, 255n34, 256n42. *See also* Land grant system; NASA; National Science Foundation; *and under individual institutions*

Gaines, Lloyd, 44. See also *Missouri Ex Rel. Gaines v. Canada*
Gaines, Matthew, 179
Gates, Robert, 15, 172–173, 176, 189–191, 202, 264n54, 269n23
Geiger, Roger, 50
Gender, 1, 3, 8, 22, 52, 95, 115, 133, 145, 172, 173, 174, 176, 177, 178, 179, 181, 190, 191, 195, 205, 206, 207, 211, 213, 225n5. *See also* Women
General Electric, 95, 103, 154, 254n31
GI Bill, 20, 49, 54
Gieryn, Thomas, 17, 116, 139
Globalization, 7, 9, 79, 142, 145–146, 181, 204, 206, 214. *See also* Capitalism; Markets
Gordon, Avery, 214
Graduate education: doctoral programs, 54, 141, 145, 160, 163, 180, 188, 196, 197, 250n68; professional programs, 11, 20, 24, 27, 30, 35, 44, 56, 67–69, 74, 75, 151, 162, 165
Graduates, engineering, 9, 33, 34, 37, 42, 50, 66, 67, 88, 94, 98, 99, 101, 102, 104, 107, 109, 117, 118, 130, 143–144, 149, 159, 162, 180, 194, 195, 216, 245n3. *See also* Employment
Grayson, Lawrence, 117–118
Greaux, Austin E., 153–154, 195
Gurin, Patricia, 164, 244n97

Hammond, David, 133
Hampton Institute, 23, 37, 182

Hansen, Arthur, 156, 158–159, 256n47, 256n49
Heald, Henry, 84–85
Health care, 28, 39, 68, 71, 114, 117, 127, 128, 158, 174
Henry, David, 84, 103
High school. *See* Secondary schools
Hines, Charles, 166–170
Hispanic Americans, 1, 2, 6, 9, 16, 82, 90, 144, 146, 168, 172, 175, 177–180, 183, 187, 189, 192, 197, 215, 216, 251n4, 255n34, 260n4
Historically black colleges and universities (HBCUs), 11, 15, 21, 24–25, 41, 42, 53, 57–58, 69, 147–149, 151, 157, 159, 160, 162–163, 164, 167–169, 174–176, 191–193, 195, 197–203, 251n3, 252n9, 252n11, 255n38, 255n41, 257n53, 259n79, 266n74
Holderman, James, 129–130
Hoover, Herbert, 65
Hopwood, Cheryl, 188
Hopwood v. Texas, 146, 167, 172, 176, 188
Housing, 28, 46, 64, 80, 82, 84–85, 88–92, 114, 116, 118, 119, 121, 127, 128, 130, 133, 238n16
Houston, 149–151, 158, 159, 162, 169, 195, 257n60, 259n80
Howard University, 5, 11, 37, 43–44, 50, 53, 154, 159, 182, 225n4, 231n15
Hrabowski, Freeman, 8, 211
Huff, Wilbur J., 33
Humanities, 23, 83, 91, 117, 127, 132, 136–137, 177, 181
Hutchison, Peyton, 65, 70
Hygiene, 39, 67, 70. *See also* Behavior, social

Identity, 10, 15, 31, 40, 91, 110, 150, 160, 172, 173, 176–177, 179, 203, 205–207, 210, 245n5; agrarian, 38–39; black, 26, 38, 69, 77, 88–89, 162–165, 168, 173, 174, 176, 177, 191–193, 196, 207; engineering, 11, 13, 16, 131–132, 135, 139, 174, 180, 207, 216, 225n5, 267n6; ethnic, 3, 6, 64, 82, 147, 179; politics of, 13, 77, 147; racial, 26, 112, 164–165, 174, 176, 191, 193, 207
Ideology, 3, 4, 5, 8, 13, 14, 16, 19, 27, 40, 66, 67, 68, 76, 81–83, 98, 130–131, 142, 147, 148, 151, 153, 163, 168, 171, 179, 181, 189, 190, 200, 205, 207, 209, 218, 236n1, 238n20, 269n23; "Curly" Byrd,

Ideology (continued)
37, 45, 58, 60; modernizing, 21, 47, 81, 169, 173; political, 9, 49, 102, 213, 223n14; racial, 12, 21–22, 28, 37, 45, 58, 60–62, 177, 179, 212, 227n15. See also Conservatism; Culture; Progressivism

Illinois, 31, 81, 87, 90, 92, 94, 120, 123, 126; Board of Higher Education (IBHE), 127–130; legislature, 90–93, 122, 127; "Regional Role" report, 135, 140

Illinois Institute of Technology (IIT), 3, 13, 80–89, 90, 91, 92, 93, 93–103, 108–111, 113, 116, 118, 120–122, 124–125, 130–131, 134, 135–137, 141, 165, 177, 209, 247n20, 247n21; Board of Trustees, 84, 96, 108; campus design, 85–87, 238n20; constituencies, 93–95; curriculum, 98–99, 102, 136–137, 141; diversification, 96–103; Early Identification Program, 96–99; facilities, 80, 85, 86, 87, 100; faculty, 87, 89, 93, 120–121, 122, 124–125; funding, 86–88, 100, 103, 111, 121, 130; as isolated, 86–89, 122–123; Metropolitan Studies Center, 125, 131; minority engineering programs, 96–101; research, 83, 85, 87–88, 94, 118–122, 124–125; urban mission of, 84–85, 88, 93

Illinois Institute of Technology Research Institute (IITRI), 87, 121

Immigration, 6, 83, 90, 92, 111

Inclusivity, 13, 52, 70, 103–106, 109, 147, 163, 167, 175, 176, 199, 201, 214

Individual ability. See Ability, individual

Industrialization, 12, 25, 26, 27, 30, 38, 40, 41, 45, 49, 80–82, 146, 149, 150. See also Capitalism; Markets

Industry: aeronautics, 52, 54, 194; aerospace, 80, 108, 117, 128, 140, 150, 194, 196, 200; automobile, 27, 158, 182, 194; computing, 80, 150, 158, 162, 194; electronics, 117, 150, 195; heavy, 80, 121; high-tech, 98, 141, 147, 150, 162, 194, 241n58; petroleum, 150, 194. See also Defense: defense industry; Funding; Military-industrial complex

Inequity. See Equity, economic

Institute of Gas Technology, 85

Integration, 26, 49, 53, 100, 113, 157; IIT, 100–101; Maryland, 12, 20, 43–45, 51–52, 56, 60–61, 71–77, 79, 228n31; Texas, 144–150, 152, 168, 175, 177.

See also Brown v. Board of Education of Topeka; Office of Civil Rights; Segregation

Interdisciplinary programs, 104–105, 119, 125, 130–131, 138, 165, 181, 183–184, 195, 197, 261n14

Jackson, Shirley Ann, 210
Jim Crow laws, 19, 49, 257n58
Johns Hopkins University, 31–33, 54, 55, 57, 232n26
Johnson, Arthur, 209–210
Johnson, Howard W., 131
Johnson, Lyndon B., 94, 152, 200, 249n64
Johnson Space Center, 150, 195, 266n74. See also NASA
Jones, Reginald, 95

Kelly, Fred J., 63
Kemper, John, 135
Kennedy, John F., 90, 94
Kerner, Otto, 94
Kim, Claire Jean, 189, 209, 217, 264n52
Kirby, Kelvin, 201, 202
Knowledge: technical vs. social, 131–139, 177, 190; universalizing, 12, 47, 134. See also under Engineering

Labor, 6–7, 9, 20, 22, 26, 34, 46, 108, 110, 181; and diversity, 7, 49, 62, 108, 110, 111, 142, 145–146, 204, 217, 224n20; labor market, 65, 110, 133, 140, 149, 150, 161, 203; wartime demand, 49, 52. See also Employment; Markets

Lamar University, 153
Land grant system, 12, 15, 20–25, 27–28, 31, 33, 35–37, 40–42, 50–51, 56–63, 76, 81, 90, 114, 119, 120, 123, 128, 144, 148–150, 158, 169, 175, 177, 179, 189, 193–194, 233n39, 235n69, 259n79. See also individual institution names

"Leadership," black, 66–69, 164
Leonard, Wilmore B., 72
Levinson, David, 107
Lewis Institute, 83–85
Light, Jennifer, 115–116, 134, 200
Lincoln University, 44, 256n49
Long, Edgar F., 72
Louisiana, 28, 146, 227n19
Louis Stokes Alliance for Minority Participation (LSAMP), 174, 180, 184. See also Alliance for Minority Participation

Loury, Glenn, 112
Lucena, Juan, 181

Malcom, Shirley, 8
Management, 134, 140, 146, 214; academic, 69, 167; engineering profession, 34, 37–38, 94, 117, 122, 136; training programs, 53, 71
Marbury, William L., 56, 60
Marbury Report, 56–57, 61
Mark, Hans, 195
Markets, 4, 33, 94, 119, 122, 129, 134, 137, 143, 146, 212–214; labor market, 65, 110, 133, 140, 149, 150, 161, 203; segregated economic spheres, 37–40, 62–65, 164. *See also* Capitalism; Globalization
Marshall, Thurgood, 21, 43, 74
Martin, Glenn L., 54–55, 57
Martin, Thomas Lyle, Jr., 96–99, 101, 103, 165
Maryland: as border state, 12, 20–22; higher education funding, 28–29, 32, 33, 35–36, 41–42, 43, 45, 50, 54–61, 66, 75, 76, 232n26, 233n31, 233n39; legislature, 12, 21, 23, 26, 27, 28, 32, 33, 35, 36, 42, 43, 50, 55, 57, 58, 59, 63, 66, 73, 74–75, 76, 233n32, 233n39; Soper and Marbury Commissions, 55–57, 60–61, 231n22; State Commission on Negro Education, 35. *See also* Baltimore
Maryland Agricultural College, 23, 27. *See also* University of Maryland, College Park
Massachusetts, 83, 150
Massachusetts Institute of Technology, 32, 37, 50, 54, 114, 121, 124, 126, 131, 178, 182
Mathematics, 2, 9, 24, 94, 97, 100, 101, 102, 104, 112, 115, 118, 133–134, 141, 146, 160, 165, 172, 179, 181, 183, 186, 188, 199, 245n2
McCready, Esther, 73
Mechanical arts, 23, 25, 29, 33, 37, 61, 108, 149, 150, 194; as distinguished from engineering, 36, 37
Mechanical engineering, 31, 32, 33, 37, 42, 52, 85, 87, 126, 130, 150–152, 158, 186, 195; as distinguished from mechanical arts, 36, 37
Media, 8, 17, 31, 43, 57, 60, 72, 73, 106, 156, 169. *See also Baltimore Sun*
Mentoring, 2, 10, 99, 153, 180

Merit, 9–10, 15, 16, 49, 65, 68, 77–78, 103, 111–112, 141, 144, 170, 171, 173, 187, 189, 190, 197–204, 206, 217, 223n14, 245n1. *See also* Eligibility
Michael Reese Hospital, 84
Migration, 40, 49, 62, 64, 84, 90, 133
Military, U.S., 8, 13, 26, 32, 48, 49, 95, 114–118, 120, 121, 124, 129, 130, 134, 135, 137, 141, 144, 147, 163, 177, 194, 200, 209. *See also* Defense
Military-industrial complex, 13, 117, 134, 135, 141
Miller, Jonson, 215
Minority engineering programs (MEPs), 4, 7, 10, 13, 15, 81, 95–101, 103, 113, 115, 146–148, 154, 170, 177–181, 184, 199, 203–204, 210, 211, 214
Mississippi, 41, 146, 225n3
Missouri Ex Rel. Gaines v. Canada, 44, 73
Mobility: economic, 36, 40, 193; professional, 95, 144, 181; social, 70
Modernization, 40, 136; academic, 12, 28, 141, 158, 177; economic, 12, 25, 30, 38, 41, 79, 80–82, 150; ideologies, 21, 47, 64, 169, 173. *See also* Development, economic; Industrialization
Morgan State College, 21, 23, 29, 35–36, 38, 55–61, 63, 66, 73, 74, 75, 76, 182, 232n24, 233n39, 236n78
Morrill Act: (of 1862), 20, 27, 148, 225n3; (of 1890), 20, 23. *See also* Land grant system
Morris, Lorenzo, 169, 251n3
Moynihan, Daniel Patrick, 134–135
Multiculturalism, 14, 76, 142, 209, 217
Murray, Donald, 20, 43, 51, 74, 76
Myrdal, Gunnar, 2

NAACP, 12, 20–21, 28–29, 43, 53–54, 57, 59, 72, 74, 148–149
NASA, 15–16, 85, 148, 150, 151, 174, 191–192, 195–203, 209, 254n31, 265n63, 266n74, 267n81; Network Resource and Training Sites (NRTs), 197–198
National Academy of Engineering (NAE), 95–96, 214; Committee on Minorities in Engineering, 95–96, 103
National Action Council on Minorities in Engineering (NACME), 1–2, 4, 8, 11, 95–96, 100, 154, 208, 222n3
National Fund for Minority Engineering Students, 95

National Institutes of Health, 121
National Science Foundation, 2, 4, 15, 48, 100, 103, 105, 121, 139, 157, 174, 176, 179–187, 211, 212, 261n15, 263n33. *See also* Alliance for Minority Participation; Engineering Education Coalitions; Foundation Coalition
National Society of Professional Engineers, 102
National Urban League, 53
National Youth Administration, 53
Native Americans, 1, 2, 6, 9, 13, 16, 82, 168, 192, 197, 216
Nelson, Ivory, 158–159
Networks, social, 5, 8, 32, 80, 84, 90, 108, 152, 153–154, 179, 185, 197–198, 263n39
New Deal, 40
Newfield, Christopher, 10, 111, 140, 223–224n14, 245n1, 247n22, 268n6
North Carolina Agricultural and Technical College, 37, 168
Northwestern University, 86, 121
Nuclear weapons, 48, 83, 114, 135, 136
Nursing, 6, 20, 22, 73, 149, 158, 194, 250n3

Obama, Barack, 9, 16, 173, 217, 269n23
Objectivity, 13, 65, 111, 131, 152
Office of Civil Rights, 144, 157–158, 159, 166, 168–169, 175–176, 191. *See also* Integration
Office of Education, 12, 51, 53, 62–71, 77, 229n1, 235n63
Ogilvie, Richard, 129
Omi, Michael, 17
Outcomes, educational, 10, 101, 133, 211
Outreach, 62, 100, 115, 127, 129, 154, 172, 192, 215, 254n28, 261n14. *See also* Minority engineering programs; Recruitment

Parker, Norman, 103, 109
Parker, Rosemary, 209–210
Participation. *See* Enrollment patterns
Pasley, Paul, 109–110
Patronage, 4–5, 8, 24, 32, 53, 65, 92, 103, 121, 126, 181, 191, 214. *See also* Funding
Pawley, Alice L., 216
Pearson, Raymond A., 28–29, 43
Pedagogy, 4, 10, 11, 13, 23–24, 28, 37–38, 101, 102, 104, 105, 107, 108, 130, 142, 167, 174, 178, 182–187, 191, 202, 210, 245n5. *See also* Curriculum, academic; Interdisciplinary programs

Pennsylvania State University, 31, 182, 233n39
Perry, Ervin S., 153
Person, Carl, 23, 232–233n29
Pickering, George, 92
Pierre, Percy A., 159–167, 175, 179, 194, 257n53
Planning, urban, 84–89, 91, 108, 116, 125, 127, 128, 132–133
Pollution. *See* Environmental issues
Porter, Theodore, 137
Poverty, 41, 46, 79, 82, 86, 88–89, 91, 92, 98, 109, 115, 125, 136, 138, 149, 156, 200, 237n5
Prairie View Agricultural & Mechanical University (PVAMU), 15, 41, 143–145, 147–170, 173–177, 179, 191–192, 194–203, 209, 250n3, 252n9, 252n15; Banneker Honors College, 160–162, 164–168, 257n54; Center for Applied Radiation Research, 195–196, 198–199, 201, 203; Engineering Concepts Institute, 154; facilities, 151, 155–157, 158, 159, 166, 175, 188, 194, 198, 202; funding, 144, 148, 151–157, 160, 166, 169, 174, 175, 192, 194–196, 199, 202, 259n79; graduate education, 153–154, 161–162, 165, 166, 194, 196; identity politics, 163–165, 168, 176, 179, 192–193, 196; as magnet, 159–162; mission, 147, 153, 162, 165, 168; and NASA, 15–16, 148, 174, 191–193, 195–203, 209, 266n74; nonminority enrollments, 162, 168, 176, 191; research, 148, 151, 153–154, 160, 164, 169–170, 173–175, 193–203, 209, 266n74; selectivity, 147, 160, 165, 167
Pre-college programs. *See* Minority engineering programs
Prestage, Jewell, 161, 163
Price, Richard, 153
Princess Anne. *See* University of Maryland, Eastern Shore
Professionalization, 80, 139, 184
Professional schools, 11, 20, 24, 27, 30, 35, 44, 56, 67–69, 74, 75, 151, 162, 165
Progressive Era, 117–118, 132
Progressivism, political, 4, 8, 11, 13, 15–16, 28, 43, 49, 51, 53, 55, 57–65, 72, 81, 89, 98, 99, 111, 114, 116, 120, 127–130, 142, 144, 145, 157, 174, 182, 183, 191, 192, 199, 201, 214, 217–218, 231n22

Quantification, 131–134, 137

Race relations, 2, 8, 11, 15, 39, 59, 60, 92, 98, 147, 168, 171, 178, 231n22
Racial conflict, 79, 89, 146; riots, 80–81, 84, 88–89
Racial preferences. *See* Affirmative action
Racism, 12, 15, 22, 38, 45, 58, 60, 146, 171, 187, 208, 209, 218. *See also* Biology (and race)
Rational Choice theory, 134
Reagan, Ronald, 14, 147, 192, 200
Recruitment, 2, 9, 15, 98, 100, 104–105, 114, 139, 144, 153, 161–162, 166–167, 172, 178, 180, 182, 187–190, 202, 211, 213, 262n16. *See also* Minority engineering programs
Regents of the University of California v. Bakke, 7, 188, 251n3
Religion, 70, 83, 190, 195
Remedial instruction, 97–99, 101–102, 106, 110, 113, 125, 141, 199, 208, 217
Reputation, institutional, 4, 27, 31, 33–34, 37, 54, 87, 90, 98, 101, 106, 129, 141, 151, 159, 167, 182, 199, 210, 211, 217
Research. *See* Agriculture; Defense; Eligibility; Foundation Coalition; *and under specific institutions*
Retention: faculty, 126, 211; minority student, 2, 99, 101, 126, 139, 172, 178, 182, 184, 185, 190, 207, 209
Rettaliata, John, 85, 87–88, 109, 134–136, 249n58
Reverse discrimination, 8, 103, 141. *See also* Affirmative action; Color blindness; Discrimination
Reward systems (academic), 102, 108, 119, 125, 138, 164, 186–187, 199, 202, 210–212, 264n44. *See also* Tenure and promotion
Richardson, Herbert, 178
Rierson, Jeanne, 178, 191
Rigor, academic, 5, 10, 18, 77, 82, 87, 99, 102, 103, 111, 132, 139, 141, 144, 164, 174, 199, 207–208, 217
Riots. *See under* Racial conflict
Robbins, Paul H., 102
Roosevelt, Franklin D., 49, 53, 61
Rose-Hulman Institute of Technology, 183, 184, 212
Rosen, George, 90
Rudder, Earl, 177, 190
Rural life, 5, 6, 25, 27–28, 36–41, 45–46, 49, 61–64, 66, 70, 90, 127, 172, 214, 233n39. *See also* Agriculture

Saline, Lindon, 103
Scholarships, 10, 15, 29, 36, 43, 100, 144, 147, 161, 167, 169, 180, 188, 189, 190, 209; out-of-state, 28, 36, 45, 72, 188
Science, and democracy, 48–49, 199. *See also* Knowledge; Objectivity; Technology
Secondary schools, 9, 11, 23–26, 42, 66, 88, 94, 97–98, 100, 104, 153, 159, 162, 167, 172, 173, 178, 180, 188, 208, 216, 217
Segregation, 3, 5, 12, 16, 19–22, 24–26, 29–30, 35–36, 39, 42–47, 49–52, 55, 58–64, 66, 68, 71–77, 79, 146, 148–149, 152, 162, 168–169, 177, 194, 208, 231n22; economic spheres, 37–40, 62–65, 164; "separate but equal" requirements, 21, 22, 28, 35, 43, 44–45, 51–52, 58, 60–61, 68, 72, 75, 146, 251n8. *See also* Discrimination; Integration; Office of Civil Rights
Selectivity, 10, 27, 77, 82, 87–88, 93–94, 99, 110, 144, 147, 165, 167, 210. *See also* Admissions; Standards
Shabazz, Amilcar, 146, 148, 150, 251n8
Shapin, Stephen, 203, 208, 267n6
Sheehy, Ronald, 160
Simon, Harold, 122
Skidmore, Owings and Merrill, 86, 91
Slaughter, John Brooks, 2, 8, 9–10, 208
Smith, William R., 97
Smith-Hughes Vocational Act (1917), 41–42
Smith-Lever Act (1914), 24, 41
Social sciences, 117, 122, 127, 132, 138
Society for the Promotion of Engineering Education, 34
Soper, Morris, 55, 231n22
Soper Commission, 55–56
South, U.S., 5, 12, 16, 19–22, 24, 26, 27, 28, 36, 38–47, 49, 52, 61, 71, 73, 76–77, 144, 148, 150, 227n15
Southern Conference Educational Fund, 73
Standards, 5, 10, 15–18, 20, 27, 28, 29, 50, 65, 77–78, 82, 96, 98, 101–102, 106, 108–110, 113, 119, 125, 139–140, 144–145, 154, 161, 172–173, 183, 191, 199, 202, 207–213, 218, 235n69, 259n76
Stanford University, 121, 182, 196
Steinberg, S. Sidney, 32–33
Stewart, R.M., 62–64, 66
Structure, social. *See under* Discrimination
Suburbs, 80–81, 84, 90, 92, 121, 132

Supreme Court, 20, 43–44, 51, 74, 146, 149, 151, 172, 176, 188, 190. *See also* specific cases
Sweatt, Heman, 149, 151
Sweatt v. Painter, 151

Teacher training, 21, 23, 25, 29, 34, 42, 54, 61, 73, 148, 250n3
Teaching. *See* Curriculum, academic; Pedagogy
Technical training, 23–25, 34, 37–42, 50, 53–54, 83, 87, 93, 102, 109, 148, 149, 151, 158, 194, 209; wartime, 49, 52–54, 71. *See also* Curriculum, academic; Engineering technology
Technology, and social knowledge, 131–139, 177, 190, 215–217. *See also under* Engineering
Tenure and promotion, 69, 102, 125, 140, 186–187, 212. *See also* Reward systems (academic)
Terrell, Glenn, 106
Tests, 5, 96, 98, 101, 104, 107, 130, 188, 218, 225n5; SAT, 154, 160, 162, 178, 208, 209
Texas: constitution, 148, 155; economy, 146, 149–150, 155, 194, 253n18; federal civil rights pressure, 157–159; Higher Education Coordinating Board, 152, 158, 160, 169, 175; Priority Plan (2000), 169; state funding, 144, 151–157, 166, 175, 188, 194, 255n34, 256n42; state legislature, 144, 154–158, 166, 188, 194; Texas Plan, 158, 162, 175; Top 10 percent law, 188–190, 264n49
Texas A&M at College Station (TAMU), 15, 144–147, 150–151, 153, 155, 157, 167, 169, 170–191, 194, 195, 196, 199, 201–202, 203, 209, 215, 254n28, 255n34, 256n49, 260n4, 261n14, 261n15, 262n16, 262n22, 264n54; Aggie spirit, 173, 175–176, 178–179, 190, 191; diversity, 175–176; Foundation Coalition, 181–187, 191, 199, 201–202, 209, 212, 263n33; funding, 175, 183, 186, 187; minority engineering programs, 147–148, 170, 177–181, 184; obstacles to reform, 186–191. *See also* Texas Plan
Texas A&M Research Foundation, 194
Texas A&M system, 15–16, 143, 146, 150, 151–152, 155, 156, 158, 166–170, 172–173, 175–177, 180, 183, 187–190, 193–194, 202, 255n34, 255n38

Texas Engineering Experiment Station, 150, 183, 194
Texas Plan, 158, 162, 175
Texas Society of Professional Engineers, 153
Texas Southern University, 149, 157–158, 162, 169, 182, 194, 257n58
Thelin, John, 25, 31, 40, 54, 177
Thomas, Alvin I, 143–145, 149, 153–154, 156, 158–159, 163–164, 173, 256n42
Thomas, Nathaniel, 97
Title VI. *See under* Civil Rights Act (1964)
Trades training. *See* Technical training
Training, technical. *See* Technical training
Transfer, 152, 162, 180, 183
Transportation, public, 46, 90, 114, 116, 118, 119, 123, 127, 128, 130, 133, 150, 213
Tuskegee Institute, 23, 41, 159, 182
Tutoring, 104, 105, 125, 178, 184

Universities (economic role), 25, 34, 77, 80, 93, 105, 114–122, 127, 134–136, 140, 145, 147, 163, 189, 194, 234n47. *See also* Development, economic; *and under* Engineering
University of Alabama, 183, 187
University of California, Berkeley, 40, 124, 182
University of California, Davis, 135
University of California, Los Angeles, 106
University of Georgia, 44
University of Houston, 157, 188, 254n28, 257n58
University of Illinois, Champaign-Urbana, 90, 92, 103, 107, 121–123, 126, 127, 130
University of Illinois, Chicago (UIC), 13–14, 80–82, 87, 89–93, 103–111, 113, 116, 118, 120, 122–123, 126–131, 133, 138, 139, 141, 177, 183, 208, 210–211, 246n17, 248n38; constituencies, 93–95; curriculum, 104–108, 113–114, 127, 129–131, 139–141, 208, 210–211; Educational Assistance Program, 103–110; facilities, 103–104, 130; funding, 108, 111, 118, 126–127, 130; interdisciplinary study, 104–105, 125, 130; relations with community, 92, 105–106, 122–123, 127, 129; research, 104, 106–107, 120, 122–123, 125–131, 141; urban mission, 120, 127–130

University of Illinois system, 103–104, 127, 129
University of Maryland, College Park, 12, 20, 21, 22–23, 27–39, 41–44, 50–51, 52, 54, 56, 58; integration, 43, 72–76, 228n31
University of Maryland, Eastern Shore, 12, 15, 20–21, 23–25, 28–30, 34–42, 45, 50–63, 66, 70, 74, 75–76, 226n6, 232n24, 232n29, 233n31, 233n39, 234n53; and agriculture, 38–42; facilities, 34–35, 50, 51, 53; faculty, 35, 37, 50, 57, 58; funding, 23–24, 28–29, 36, 42–43, 45, 50, 56, 58–61, 66, 76–77; investment in, 50–52, 55–57, 60, 233n31; as land-grant institution, 23–27, 66; race relations, 39, 59, 60; research, 25, 37, 61, 75; technical instruction, 34–38. *See also* Morgan State College
University of Maryland system, 12, 15, 19–22, 28–31, 35–36, 39–41, 44–45, 50–51, 53, 55–59, 61, 76–79, 82; Board of Regents, 28, 43, 50, 74–76; as dual educational system, 21, 28, 30, 51, 59, 61, 76; and integration, 72–76
University of Michigan, 124, 172, 176
University of Missouri, 44, 73
University of Texas, 44, 146, 148–149, 153, 155, 157–158, 172, 187, 188, 190, 194, 253n20, 254n29
University of Wisconsin, 95, 114, 263n33
University v. Murray, 20, 43, 51, 74, 76
Urban, Wayne, 77
Urban decay, 81, 84–91, 93, 115, 117–119, 185, 200
Urban development, 3, 38, 80–95, 103, 109, 114–134, 200, 238n16, 248n38. *See also* Poverty

Values, social, 39, 40, 44, 46, 67, 125–126, 138, 179, 191, 199, 200, 209; democratic, 8, 49; in engineering, 14, 119, 125, 138, 191; racial, 11, 21, 51
Van der Rohe, Ludwig Mies, 85
Vandiver, Frank E., 178
Vietnam War, 13, 83, 114, 129, 134, 137, 138, 139, 247n29
Virginia, 189, 225n3
Vocational training. *See* Technical training

Washington, Booker T., 23
Watkins, William H., 22, 218, 228n39
Watson, Karan, 186, 187, 264n44
Weinberger, Caspar, 200
West Virginia, 28
Wharton, David, 37
White, Mark, 157
Whittle, Hiram, 73–76
Wilkins, Richard, 201, 202
Williams, Aubrey, 53
Williams, Jack K., 177
Williams, John T., 57–60, 63, 232–233n29, 233n32, 233n39
Wilson, C. L., 151
Wilson, Ron, 155
Winant, Howard, 17, 246n16
Wisnioski, Matthew, 124
Women, 1, 3, 7, 52, 105, 110, 177, 186, 202, 204, 216. *See also* Gender
Wood, Peter, 216
Workforce. *See* Labor
World War I, 25, 31, 39, 64
World War II, 5, 7, 12, 13, 20, 22, 24, 25, 31, 42, 45, 48–54, 58, 68, 71, 77, 80, 87, 90, 93–94, 121, 133, 146, 150, 151, 229n1, 231n15, 231n22, 253n20
Wright, George, 169
Wulf, William, 214

Young, Michael, 99, 242n69

Harvard University Press is a member of Green Press Initiative (greenpressinitiative.org), a nonprofit organization working to help publishers and printers increase their use of recycled paper and decrease their use of fiber derived from endangered forests. This book was printed on 100% recycled paper containing 50% post-consumer waste and processed chlorine free.